Simone Hamerla

Interaction Quenches in Fermionic Hubbard Models

Simone Hamerla

Interaction Quenches in Fermionic Hubbard Models

Dynamics of Quenched Fermions in One and Two Dimensions

Südwestdeutscher Verlag für Hochschulschriften

Impressum / Imprint

Bibliografische Information der Deutschen Nationalbibliothek: Die Deutsche Nationalbibliothek verzeichnet diese Publikation in der Deutschen Nationalbibliografie; detaillierte bibliografische Daten sind im Internet über http://dnb.d-nb.de abrufbar.

Alle in diesem Buch genannten Marken und Produktnamen unterliegen warenzeichen-, marken- oder patentrechtlichem Schutz bzw. sind Warenzeichen oder eingetragene Warenzeichen der jeweiligen Inhaber. Die Wiedergabe von Marken, Produktnamen, Gebrauchsnamen, Handelsnamen, Warenbezeichnungen u.s.w. in diesem Werk berechtigt auch ohne besondere Kennzeichnung nicht zu der Annahme, dass solche Namen im Sinne der Warenzeichen- und Markenschutzgesetzgebung als frei zu betrachten wären und daher von jedermann benutzt werden dürften.

Bibliographic information published by the Deutsche Nationalbibliothek: The Deutsche Nationalbibliothek lists this publication in the Deutsche Nationalbibliografie; detailed bibliographic data are available in the Internet at http://dnb.d-nb.de.
Any brand names and product names mentioned in this book are subject to trademark, brand or patent protection and are trademarks or registered trademarks of their respective holders. The use of brand names, product names, common names, trade names, product descriptions etc. even without a particular marking in this works is in no way to be construed to mean that such names may be regarded as unrestricted in respect of trademark and brand protection legislation and could thus be used by anyone.

Coverbild / Cover image: www.ingimage.com

Verlag / Publisher:
Südwestdeutscher Verlag für Hochschulschriften
ist ein Imprint der / is a trademark of
OmniScriptum GmbH & Co. KG
Heinrich-Böcking-Str. 6-8, 66121 Saarbrücken, Deutschland / Germany
Email: info@svh-verlag.de

Herstellung: siehe letzte Seite /
Printed at: see last page
ISBN: 978-3-8381-3538-0

Zugl. / Approved by: Dortmund, TU Dortmund, Diss.,2013

Copyright © 2014 OmniScriptum GmbH & Co. KG
Alle Rechte vorbehalten. / All rights reserved. Saarbrücken 2014

Meinem geliebten Papa Andreas Hamerla

Contents

1. **Abstract** — 7

2. **Introduction** — 11
   - 2.1. Non-Equilibrium — 11
   - 2.2. Theoretical Concepts — 13
   - 2.3. Methods — 16

3. **Model and Methods** — 19
   - 3.1. The Fermionic Hubbard Model — 19
     - 3.1.1. Issues in Non-Equilibrium — 22
   - 3.2. Iterated Equation of Motion Approach — 24
     - 3.2.1. Normal Ordering — 28
     - 3.2.2. Convergence — 31
   - 3.3. Implementation — 36
   - 3.4. Self-Consistent Truncation — 39
   - 3.5. Runaway Time — 46

4. **Variants of the Approach** — 51
   - 4.1. Matrix Approaches — 51
     - 4.1.1. Non-unitarity — 52
     - 4.1.2. Different Scalar Products — 52
   - 4.2. Momentum-Space Approach — 57
   - 4.3. Self-Similar Calculations in U^2 — 64
     - 4.3.1. Non-Interacting Case — 64
     - 4.3.2. Interacting Case — 66
     - 4.3.3. Comparison to the 11-Loop Calculation — 76

Contents

5. Results for the One-Dimensional Model — 77
- 5.1. Half-Filled Model — 77
  - 5.1.1. Jump for Various Interaction Strengths U — 77
  - 5.1.2. Momentum Distribution — 80
- 5.2. Influence of Doping — 86
- 5.3. Bosonization — 94
  - 5.3.1. Spinless Fermion Model — 95
  - 5.3.2. General Concepts — 99
  - 5.3.3. Results for Spinless Fermions — 101
  - 5.3.4. Bosonization in the Quarter-Filled Hubbard Model — 110
  - 5.3.5. Relevance of the Results for Longer Times — 119
- 5.4. Periodicity — 121
  - 5.4.1. Strong Quenches — 121
  - 5.4.2. Dynamical Transition — 126

6. Two-dimensional Model — 133
- 6.1. Half-Filled Two-Dimensional Model — 134
  - 6.1.1. Convergence — 137
  - 6.1.2. Results for the Half-Filled Two-Dimensional Model — 138
  - 6.1.3. Strong Quenches — 140
  - 6.1.4. Weak Quenches — 143
  - 6.1.5. Strict U^2–Calculations — 143
  - 6.1.6. Momentum Dependence of the Jump — 149
  - 6.1.7. Full Momentum Distribution — 154
- 6.2. Doped Two-Dimensional Model — 156
  - 6.2.1. Second Order Results away from Half-Filling — 160
  - 6.2.2. Momentum Distribution — 163

7. Summary and Outlook — 165
- 7.1. Summary — 165
- 7.2. Outlook — 167

Contents

Appendix   169

A. Second Order Calculations   **171**
   A.1. Calculation of the Commutator 171
   A.2. Determination of the Spectral Density 174

B. Other Truncation Schemes   **179**
   B.1. Truncation According to Order in U and J 179
   B.2. Omin-Omax Truncation 181

C. Determination of the Luttinger Parameters   **185**

D. Comparison of the Two-Dimensional and the Infinite-Dimensional Model   **193**

Bibliography   **195**

Contents

1. Abstract

In the last years the impressive progress on the experimental side led to a variety of new experiments allowing to address systems out of equilibrium. In this way the behavior of such systems far from equilibrium is no longer a purely theoretical issue but indeed observable. New experimental techniques, like particles trapped in optical lattices, render a realization of quantum systems with nearly arbitrary system parameters possible and provide a possibility to study their time evolution. Systems out of equilibrium are characterized by the fact, that these systems are in highly excited states giving rise to totally new fascinating properties.

In the present thesis one- and two-dimensional fermionic Hubbard models out of equilibrium are discussed. The system is taken out of equilibrium by a so-called interaction quench. At the beginning the system is prepared in the groundstate of the non-interacting Hamiltonian. At a time t the interaction between the fermions is suddenly turned on so that the time evolution is governed by the whole, interacting Hamiltonian. Hence the system is prepared in the groundstate of one Hamiltonian but evolves according to a different Hamiltonian. Consequently the system ends up in a highly excited state.

To describe such a system a method based on an expansion of the Heisenberg equations of motion to highest order possible is developed in this thesis. This method provides an exact description of the time evolution on short and intermediate time scales after the quench. As the method reveals exact results and does not rely on any perturbative assumption, a study of arbitrarily large interaction strengths is possible. Besides, the method is one of the few methods capable of two-dimensional systems.

In the following the method used in this thesis is explained and advantages and disadvantages of the approach are thematized. For this purpose the results of the developed iterated equation of mo-

Abstract

tion approach are compared to results obtained in a calculation up to second order in the interaction. Then the dynamics of the one-dimensional model is discussed with a focus on the relation of the results derived by the iterated equations of motion approach to results obtained by bosonization theory, the behavior for strong interactions and the dynamical transition from the weak to the strong quench regime.

Furthermore, the Hubbard model is studied on a two-dimensional square lattice. This model is fundamentally different from the one-dimensional model: In contrast to the one-dimensional model the two-dimensional model is not integrable allowing a true relaxation of the system. For this system a calculation up to second order in the interaction is performed and compared to the results of the iterated equation of motion approach. Besides, the time evolution of the momentum distribution and the influence of doping on the dynamics is studied. Moreover, a first estimate for relaxation times is provided without relying on the assumption of a mixed state.

Zusammenfassung

In den letzten Jahren wurden vielfältige neue Experimente entwickelt, die es erlauben, Systeme jenseits des Gleichgewichtes zu betrachten. Dadurch ist das Verhalten dieser Systeme im Nicht-Gleichgewicht nicht länger eine rein theoretische Fragestellung sondern tatsächlich beobachtbar. Neue experimentelle Techniken, wie Teilchen in optischen Gittern, ermöglichen es Systeme mit nahezu beliebigen Systemparametern zu realisieren und deren Zeitentwicklung zu studieren. Systeme jenseits des Gleichgewichts zeichnen sich dadurch aus, dass sich diese in hochangeregten Zuständen befinden, wodurch sich völlig neue, faszinierende Eigenschaften ergeben.

In der vorliegenden Arbeit werden Hubbard-Modelle in ein und zwei Dimensionen jenseits des Gleichgewichtes betrachtet. Das System wird dabei durch einen sogenannten Wechselwirkungs-Quench aus dem Gleichgewicht gebracht. Zunächst befindet sich das System im Grundzustand des wechselwirkungsfreien Hamiltonoperators. Zur Zeit t wird die Wechselwirkung zwischen den Fermionen plötzlich eingeschaltet, sodass die Zeitentwicklung durch den gesamten, wechselwirkenden Hamiltonoperator bestimmt wird. Somit ist das System im Grundzustand eines Hamiltonoperators präpariert, entwickelt sich jedoch gemäß eines anderen. Dadurch wird das System in einen hochangeregten Zustand versetzt.

Zur Beschreibung eines solchen Systems wird in dieser Arbeit eine Methode basierend auf einer Entwicklung der Heisenbergschen Bewegungsgleichung zu möglichst hoher Ordnung entwickelt. Die Methode erlaubt es, die Zeitentwicklung des Systems auf kurzen und mittleren Zeitskalen nach dem Quench exakt zu beschreiben. Da die Methode exakte Ergebnisse liefert und ihr keinerlei störungstheoretische Annahmen zu Grunde liegen, ist es möglich beliebige Wechselwirkungsstärken zu betrachten.

Abstract

Außerdem ist diese Methode eine der wenigen Methoden, die auf zweidimensionale Systeme anwendbar sind.

Im Folgenden wird zunächst die genutzte Methode erläutert und deren Vor- und Nachteile diskutiert. Hierzu werden die Ergebnisse mit denen einer Rechnung bis zur quadratischen Ordnung in der Wechselwirkung U verglichen. Anschließend wird die Dynamik des eindimensionalen Modells diskutiert. Dabei wird der Bezug der Ergebnisse zur Bosonisierung, das Verhalten für besonders große U sowie der dynamische Übergang zwischen schwachen und starken Quenches thematisiert.

Des Weiteren wird das Modell auf einem zweidimensionalen Gitter betrachtet. Dieses System unterscheidet sich grundsätzlich von dem eindimensionalen System: Im Gegensatz zum eindimensionalen System ist das zweidimensionale System nicht integrabel, sodass eine Relaxation möglich ist. Auch für dieses System wird eine Rechnung zweiter Ordnung in U durchgeführt. Außerdem wird die Zeitentwicklung der Impulsverteilung betrachtet sowie der Einfluss der Dotierung auf die Dynamik studiert. Darüberhinaus wird eine erste Abschätzung der Relaxationszeiten gegeben ohne auf die Annahme eines Gemisches zu vertrauen.

2. Introduction

2.1. Non-Equilibrium

Non-equilibrium systems are of special interest as they reveal totally new fascinating phenomena such as dynamical transitions, where already small changes of the system parameters change the behavior of the system qualitatively. Systems far from equilibrium have the distinction that these are usually in highly excited states. This poses new challenges to a theoretical description. Many methods used in equilibrium situations are based on the assumption that the system is in its groundstate or at least in a state close to the groundstate. Under these conditions it is possible to map a complex system to a simpler one devised to explain the low-energy physics of the model. Often a quasi-particle based description is possible, where the groundstate of the system is regarded as a vacuum and its excitations as quasi-particles. These systems involve only a small number of quasi-particles. Due to the small density of quasi-particles the accessible phase space is drastically reduced. In contrast to this, non-equilibrium systems are highly excited exploring a much larger phase space. Consequently a description based on quasi-particles is not possible and many equilibrium methods fail.

Far from equilibrium the dynamics is governed by processes on many different energy scales. Thus a method has to cope with many degrees of freedom developing on short time scales, which further hampers the theoretical descripton.

Recently refined experimental techniques based on optical lattices [1, 2] and femtosecond spectroscopy [3] allow to study quantum systems far from equilibrium, so that non-equilibrium physics is no longer a purely theoretical playground. The possibility to change intrinsic system parameters and observe the evolution permits a totally new class of experiments (for a review see Ref. [4]).

Introduction

Optical lattices allow to artificially create lattice structures with nearly arbitrary properties and lattice geometry [5]. In 1998 Anderson and Kasevich trapped atoms in such a lattice for the first time [6]. Since then an impressive progress concerning the controllability of the system parameters such as the lattice depths has been observed. Due to this progress it is now possible to simulate given model Hamiltonians [7]. Thus non-equilibrium systems are no longer just a theoretical perspective but indeed observable.

Greiner *et al.*, for instance, used optical lattices to realize a Mott insulator made of bosons [8]. In this experiment ^{87}Rb atoms from a Bose-Einstein condensate are loaded into the lattice. On changing the lattice depths they observed a transition from a superfluid phase with weakly interacting particles to the Mott insulating phase. The transition was observed through the loss of the phase coherence in the insulating phase.

Optical lattices embody a nearly perfect decoupling from their environment, so that these systems can be assumed to be completely isolated. Thus one main source for decoherence is excluded in these experiments, allowing to observe the dynamics over fairly long times. Combining an unprecedented controllability of the system parameters with long observation times, optical lattices are an ideal testbed for non-equilibrium dynamics.

The long observation times rendered a study of collapse and revival behavior in optical lattice systems [1] possible. In this experiment the matter waves of Bose-Einstein condensates trapped in optical lattices are studied by the use of their interference patterns. After collapsing completely the interference pattern is recovered for later times. This behavior is referred to as collapse and revival. Optical lattice setups have also been used to realize a one-dimensional Bose gas and to study the influence of integrability on the dynamics of the system [2]. In comparison to these experiments, experiments with fermions are much more demanding, as the required temperatures are much harder to reach in these systems. However fermions can also be cooled and loaded into the lattice structures [9, 10]. Experiments with fermions address the control of the filling factor [11], transport processes caused by interactions and long-range order in

a condensate of fermion pairs [12]. Another study focusses on doublons in the Fermi-Hubbard model [13].
Besides optical lattice setups also other techniques have been developed to study systems far from equilibrium. Amongst these are time-resolved pump-probe experiments [14]. The pulses used in this context are shorter than the time scales of relaxation and dephasing, so that snapshots of the time evolution are feasible [5].

The possibility of tuning the intrinsic parameters over a wide range enables the realization of systems far from equilibrium. One efficient way to take the system out of equilibrium is to switch on system parameters such as the interaction strength between the particles abruptly. This scenario is known as *quench*. The quench can be restricted to local interactions [15, 16] or it can be global. In this thesis global quenches are discussed, where the interaction between the particles is suddenly turned on, i.e., *interaction quenches*. Initially the system is prepared in the groundstate of a non-interacting Hamiltonian. At a specific time t the interaction is suddenly turned on and the system evolves according to the full Hamiltonian. As the system is prepared in the groundstate of one Hamiltonian but its time evolution is governed by a different Hamiltonian it ends up in a highly excited state far from equilibrium. As the initial states are ground states of the non-interacting Hamiltonian interaction quenches allow to observe the build-up of correlations.

2.2. Theoretical Concepts

Systems far from equilibrium are in highly excited states which leads to the varied observed phenomena and at the same time adds to the appeal of these systems. Due to the high energies many fast developing degrees of freedom have to be considered. In a system taken out of equilibrium a main question concerns the thermalization of the system. In classical mechanics a well established concept is ergodicity, which explains thermalization processes. The trajectory of the respective model covers the whole energy surface in phase

Introduction

space so that the information about the initial state is lost. In this case the time average of observables equals their configuration average and the system is said to be thermalized.

In these systems the role of integrability is clearly defined. Following the Liouville formalism an integrable system contains a macroscopic number of constants of motion. As knowledge about the initial state is preserved by the constants of motion the accessible phase space is restricted and thermalization is not possible.

For quantum systems there is no general concept about ergodicity and it is unclear how a closed system relaxes at all. As the time dependence is governed by the Schrödinger equation a pure state remains a pure state over the whole time. Consequently relaxation to a phase described by stationary density matrices is not possible. However, a subsystem of a given system may still thermalize [17, 18]. Thermalization in quantum systems is defined less strictly. A quantum system is regarded as thermalized if some of its observables, for example local observables [19], take values corresponding to the ones in a thermal state. It was shown that such a thermalization is observable in quantum systems [20].

One concept used in non-equilibrium situations is the *eigenstate thermalization hypothesis* (ETH), which states the equality of the expectation value in a single energy state and its microcanonical average [21, 22]. The ETH predicts that the long time average is given by the microcanonical average, as observed [23,24]. Recently the ETH has been generalized to integrable models. In such systems local observables can be determined by considering a single representative eigenstate [25].

The effect of integrability on the time evolution of these systems is still highly debated. There are studies which revealed the lack of thermal behavior even for non-integrable systems [26, 27], while the observables in other systems are described by thermal values [28, 29]. In this context integrable Ising-type models [30, 31], spinless fermions [26] as well as spinful models [32] are discussed. The influence of breaking integrability is studied [33]. Kinoshita *et al.* observed experimentally that hard-core bosons do not show thermalization [2], which was explained by the integrability of hard-core bosons in one dimension [28]. Roux attributed the absence of

2.2 Theoretical Concepts

thermalization to finite-size effects of one dimensional systems [34].

In integrable systems memory of the initial state is preserved during the time evolution due to the constants of motion. To meet the constraints imposed by the integrals of motion Rigol introduced the generalized Gibbs ensemble with a density matrix ρ_{GGE} [28]. This approach is based on maximizing the entropy [35], while conserving the integrals of motion I via

$$\text{Tr}\left[\rho_{GGE}\hat{I}\right] = \langle\hat{I}\rangle_0 \tag{2.1}$$

$$\rho_{GGE} = \frac{e^{-\sum_m \lambda_m \hat{I}_m}}{\text{Tr}\, e^{-\sum_m \lambda_m \hat{I}_m}} \tag{2.2}$$

with $\langle ..\rangle_0$ denoting the expectation value in the initial state and λ_m denoting Lagrange multipliers. This concept has been applied successfully to spinless fermions [26] and the sine-Gordon model [36] as well as to Luttinger liquids [37]. Although the GGE has been proven to be very powerful in the description of non-equilibrium systems [38,39] its general validity is still debated [40]. It was observed that the relaxation to the GGE can be precluded by effects of disorder [41]. Thus the influence of local and non-local conserved quantities has to be studied further [42].

Besides the long time behavior, also the behavior on shorter times after the quench reveals fascinating properties such as the *prethermalization*. This term was introduced by Berges et al. [43] and denotes a time regime in which the system reaches a quasistationary non-thermal state characterized by the observation that some observables are already thermalized while others are not [44]. This regime is followed by an ensuing thermalization on considerably longer times. Such prethermalization phenomena were found in several models [44–47]. In integrable models the non-thermal states observed after the quench can be understood as the system being trapped in prethermalized states so that a further decay is prevented [48].

The present thesis focusses on short and intermediate times after the quench in the integrable one-dimensional Hubbard model and in the non-integrable two-dimensional model. In contrast to the integrable model the latter one shows a true relaxation (see Sect. 6). On the other hand, striking similarities between the two models reveal fascinating properties (see for instance the pronounced oscillations for strong quenches discussed in Sect. 5.4.1).

2.3. Methods

The impressive progress on the experimental side triggered extensive theoretical studies based on a large variety of analytical and numerical tools, for a review see Ref. [49]. As processes on many different energy scales have to be considered new theoretical methods are called for. Amongst these is an extension of the density matrix renormalization group (DMRG) technique [50, 51] based on Trotter-decomposition [52]. This extension was successfully applied to the Bose-Hubbard model [53] and ultracold fermions in optical lattices [54]. This method is, however, restricted to one-dimensional systems and not too long times. Numerical renormalization group techniques have also been developed to cope with time dependent systems [55]. Moreover, the dynamical mean-field theory was extended to describe non-equilibrium systems by the use of time dependent two-time Green functions [56–58]. This approach was applied to many systems such as the Falicov-Kimbal model [59] or the Hubbard model [45]. It is exact in infinite dimensions. Moeckel and Kehrein used a forward-backward continuous unitary transformation (CUT) [47]. The forward-backward CUT successfully described the second order of U results for the dynamics of the Hubbard model, the sine-Gordon model [60] and the Kondo model [61]. When describing higher order processes additional secular terms created during the calculation would have to be considered. These terms lead to unreasonable results. In second order in the interaction U these terms are negligible, but to achieve higher order results these are essential, so that the description with this method breaks down.

One of the few methods capable to describe two-dimensional systems is exact diagonalization [23]. This method is severly limited in system size and the calculations are demanding. Another method

2.3 Methods

used to describe two-dimensional systems is quantum Monte Carlo (QMC) [62], which was applied to the Hubbard model [63]. In this study the interaction between the particles is suddenly turned off. Thus the final Hamiltonian is bilinear which simplifies the calculations since the equations of motion are easily derived. After the quench the system is governed by its initial correlations.

In the approach presented here the quench is performed the other way round, so that the initial state is rather simple while the final state is complicated. In a quench like this the correlations are build-up with time. The approach used in this context is an expansion of the Heisenberg equations of motion to highest order possible. This approach is distinguished by its flexibility. It is capable of arbitrary interaction strengths, arbitrary fillings and different Hamiltonians. The approach developed in this thesis is used to describe the one-dimensional as well as the two-dimensional model with arbitrary quench strengths. Thus this approach complements the existing methods and allows to study unexplored properties.

Introduction

3. Model and Methods

3.1. The Fermionic Hubbard Model

The Hubbard model is one of the prototypic models for the description of interacting electron systems as it contains the hopping of the electrons as well as an interaction between them. The model was introduced in 1963 independently by Hubbard [64], Gutzwiller [65], and Kanamori [66]. Originally designed to explain ferromagnetism in transition metals the Hubbard model experienced a wide range of application such as Mott metal insulator transitions [67, 68] and high-T_C cuprates [69, 70]. In the last years a new field of application was added by the progress in experiments with optical lattices [1, 8]. Due to this progress, it is now possible to realize the Hubbard model in nearly perfect isolation from the environment and to study non-equilibrium processes in this setup.

In the Hubbard model the long ranged Coulomb repulsion of the original many-body Hamiltonian is reduced to a local interaction. This simplification is justified by screening. A sizeable density of states at the Fermi level leads to screening of the long ranged Coulomb potential between the electrons, so that the only non-negligible term is given by a pure on-site repulsion. Besides this simplification the model bases on the assumption that the electrons are restricted to one band.

For the quench, the interaction term of the Hamiltonian is abruptly switched on. Thus the Hubbard Hamiltonian becomes time

Model and Methods

dependent and reads

$$\hat{H} = \hat{H}_0 + \Theta(t)\hat{H}_{\text{int}} \tag{3.1a}$$

$$\hat{H}_0 = -J \sum_{<i,j>,\sigma} \left(\hat{c}^\dagger_{i,\sigma} \hat{c}_{j,\sigma} + \text{h.c.} \right) \tag{3.1b}$$

$$\hat{H}_{\text{int}} = U \sum_i \left(\hat{n}_{i,\uparrow} - \frac{1}{2} \right) \left(\hat{n}_{i,\downarrow} - \frac{1}{2} \right) \tag{3.1c}$$

with the operator $\hat{c}^\dagger_{i,\sigma}$ ($\hat{c}_{i,\sigma}$) creating (annihilating) a fermion with spin σ on site i of the lattice and the particle number operator $\hat{n}_{i,\sigma}$. The first part $\hat{H}_0$ describes the hopping of an electron with spin σ from site i to site j and vice versa. For this hopping to take place, i and j have to be nearest neighbors as indicated by the bracket under the sum. The corresponding hopping element is denoted by J. On a lattice with coordination number z the bandwidth is then given as $W = 2zJ$. In this thesis the Hubbard model is considered on a one-dimensional linear chain and on a two-dimensional square lattice. For the one-dimensional model the coordination number is $z = 2$ and in the two-dimensional square lattice it takes the value $z = 4$. Throughout this thesis the bandwidth is used as natural energy scale. Consequently the time is measured in the inverse bandwidth $\frac{1}{W}$. Furthermore, $\hbar$ is set to unity $\hbar = 1$. The interaction term $\hat{H}_{\text{int}}$ consists of a pure onsite repulsion with the interaction parameter $U \geq 0$. Thus putting two electrons with opposite spin on the same site i requires the energy U. Despite its simple form containing only two parameters (J and U) no general solution for the Hubbard model exists. The one-dimensional model is a special case as the model is integrable and thus solvable by Bethe ansatz [71].

In this thesis interaction quenches in the Hubbard model are studied. The quench is realized by the following protocol. At $t < 0$ the system is initialized in the groundstate of the non-interacting Hamiltonian $\hat{H}_0$. Thus interaction-free Fermi seas are used as initial states. At $t = 0$ the interaction is suddenly switched on as indicated by the Heaviside function. From this time on the system evolves

3.1 The Fermionic Hubbard Model

under the interacting Hamiltonian $\hat{H} = \hat{H}_0 + \hat{H}_{\text{int}}$. Thus the system is taken out of equilibrium giving rise to totally new challenges for a theoretical description of these systems.

Model and Methods

3.1.1. Issues in Non-Equilibrium

The initial state is the Fermi sea which exhibits a jump at the Fermi surface. In the one-dimensional case the jump is defined by

$$\Delta n(t) := \lim_{k \to k_F^-} n_t(k) - \lim_{k \to k_F^+} n_t(k) \quad (3.2)$$

with the Fermi vector k_F.
Exposed to the quench the jump is reduced with time t. A schematic sketch of the behavior of the jump can be found in Fig. 3.1. For $t = 0$ the momentum distribution jumps from $n_0(k < k_F) = 1$ to $n_0(k > k_F) = 0$. Under the influence of the quench the jump is reduced. In a thermalized system the jump vanishes completely.

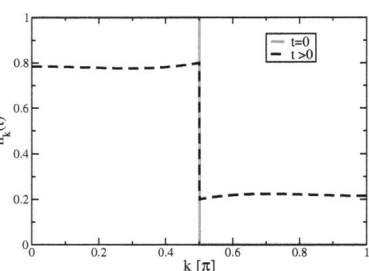

Fig. 3.1.: Schematic sketch of the jump in the momentum distribution for the initial state ($t = 0$) and after the quench ($t > 0$).

Consequently the jump can be used to probe the dynamics of the system towards a steady state. For infinite-dimensional systems the momentum distribution develops towards a momentum distribution with finite temperature $n(k)(t \to \infty) \approx n(k, T > 0)$ [48].

In a quenched model the question arises how the jump is decreased on short and intermediate times after the quench before it reaches its steady state. For the one-dimensional Tomonaga-Luttinger model field theory revealed a jump which decreases according to a power law [36, 37, 72]. On the other hand, Moeckel and Kehrein studied the Hubbard model in dimensions larger than one by the use of forward-backward CUTs. They calculated an explicit formula for the infinite-dimensional model [46, 47]. This approach yields results correct up to second order in U. From their results they concluded that fourth order processes, which are neglected in their approach, lead to an abruptly vanishing jump, due to finite life times of the fermions at the Fermi surface.

For infinite dimensions non-equilibrium DMFT was applied

3.1 The Fermionic Hubbard Model

to the Hubbard model [45]. In this study the jump decreased quickly but steadily. For this model prethermalization plateaus were observed. It was assumed that there is a difference in the relaxation of *momentum mode averaged* and *momentum mode* quantities [44]. Momentum mode averaged quantities relax already on short time scales due to dephasing effects. In contrast to this momentum mode quantities relax on later time scales. For the analytical solvable Tomonaga-Luttinger model Uhrig discussed the jump by applying a Fourier transform to the one-particle Green function. This calculation was analytically possible via a Bogoliubov transformation. Within this approach a jump decreasing according to a power law was observed. In view of these contradictory results, suggesting either a slowly decreasing or an abruptly vanishing jump, the question arises how the jump in the fermionic Hubbard model behaves. The Hubbard model is a true many particle model, as it contains interactions between the fermions. For the one-dimensional model the question arises in how far the dynamics is influenced by its integrability. Concerning this question a study of the non-integrable two-dimensional model will reveal similarities and differences giving insight into the role of conserved quantities. Besides it is unclear in how far the results for the Tomonaga-Luttinger model are relevant for lattice models like the ones discussed in this thesis. Thus the quench in the Hubbard model and its relation to the results obtained by bosonization theory are discussed.

Throughout this thesis the jump and the whole momentum distribution are used as sensitive probe for the relaxation of the model.

3.2. Iterated Equation of Motion Approach

A method describing systems far from equilibrium has to face two requirements. On the one hand, it has to capture the interaction between the particles correctly and on the other hand it has to describe the time evolution of the system. These requirements involve dealing with many degrees of freedom, which develop on short time scales. To ensure that both aspects are fulfilled a systematically controlled expansion of the Heisenberg equation of motion for $\hat{c}^\dagger_{j,\sigma}$ is introduced.

The equations of motion are expanded to the highest order possible, where the maximal order is limited by computational resources. Results obtained by this approach go far beyond the second order in U results obtained by forward-backward transformations [46, 47]. The method consists of two parts. First, the equations of motion are set up and then these differential equations are solved numerically. Although the derivation of the differential equations is performed on a computer the setting up of the differential equations is completely analytical. Thus the approach is a semi-analytical one.

Due to translational invariance the results obtained are directly the ones for an infinite lattice and there are no finite size effects.

The momentum distribution is calculated by Fourier transformation of the one-particle correlation function

$$G_\sigma(\vec{r},t) = \langle 0|\hat{c}_{\vec{r},\sigma}(t)\hat{c}^\dagger_{0,\sigma}(t)|0\rangle \tag{3.3}$$

with $|0\rangle$ denoting the initial state, i.e., the Fermi sea. Obviously the time dependence of the creation and annihilation operators $\hat{c}^\dagger$ and $\hat{c}$ is required.

To capture the time dependence the operator $\hat{c}^\dagger_{\vec{r},\uparrow}(t)$ is expanded in an operator basis. At the beginning ($t = 0$) the time dependent operator is given by a single creation operator $\hat{c}^\dagger_{\vec{r},\uparrow}(t = 0) = \hat{c}^\dagger_{\vec{r},\uparrow}$. Due to the quench to an interacting system particle-hole pairs (denoted by $T^\dagger L^\dagger$) are gradually created for $t \neq 0$. This leads to an ansatz for

3.2 Iterated Equation of Motion Approach

$\hat{c}^\dagger_{\vec{r},\uparrow}(t)$, which can formally be written as [72]

$$\hat{c}^\dagger_{\vec{r},\uparrow}(t) = \hat{T}^\dagger_{\vec{r}} + \left(\hat{T}^\dagger\hat{T}^\dagger\hat{L}^\dagger\right)_{\vec{r}} + \ldots \quad (3.4)$$

with $\hat{T}^\dagger$ ($\hat{L}^\dagger$) denoting a general superposition of particle (hole) creation operators. The particle creation operator reads

$$\hat{T}^\dagger_{\vec{r}} = \sum_{|\vec{\delta}| \lesssim v_{\max}t} \sum_\sigma h_0(\vec{\delta},t) \hat{c}^\dagger_{\vec{r}+\vec{\delta},\sigma} \quad . \quad (3.5)$$

In this ansatz the operators $\hat{c}^\dagger_{\vec{r}+\vec{\delta},\sigma}$ are summed weighted with the time dependent prefactors $h_0(\vec{\delta},t)$. The shift $\vec{\delta}$ of the operators is bounded from above by $v_{\max}t$, with $v_{\max}$ denoting the maximal velocity governed by the Lieb-Robinson theorem [73]. The velocity $v_{\max}$ is the maximal velocity of quasi particles created by the quench, thus the bound $v_{\max}t$ describes the maximal distance the excitations can travel within a time t. This bound defines a lightcone [74] containing the essential correlations. Correlations outside the lightcone are exponentially suppressed and can thus be neglected [75].

Thus to capture all processes relevant up to a certain time t all processes with a spatial range given by $v_{\max}t$ have to be considered. As such a maximal velocity is expected to appear in all dimensions the lightcone assumption does not constrain the applicability of the approach. In fact it has been observed in various models [76–78]. Starting with a single creation operator a particle-hole pair like

$$(T^\dagger T^\dagger L^\dagger)_{\vec{r}} = \sum_{\{\sigma_1,\sigma_2,\sigma_3\}} \sum_{\mathrm{dist}(\vec{\delta}_1,\vec{\delta}_2,\vec{\delta}_3) \lesssim v_{\max}t} h_j(\vec{\delta}_1,\vec{\delta}_2,\vec{\delta}_3,t) : \hat{c}^\dagger_{\vec{r}+\vec{\delta}_1,\sigma_1} \hat{c}^\dagger_{\vec{r}+\vec{\delta}_2,\sigma_2} \hat{c}_{\vec{r}+\vec{\delta}_3,\sigma_3} : \quad (3.6)$$

is created due to the quench. The index j labels the prefactors belonging to different spin configurations $\{\sigma_1,\sigma_2,\sigma_3\}$. Thus the prefactor of a term with $\{\sigma_1=\uparrow,\sigma_2=\downarrow,\sigma_3=\downarrow\}$ carries a different index than the

Model and Methods

one for a term with all spins pointing upwards. In the above formula $dist(\delta_1, \delta_2, \delta_3)$ denotes the distance of the term from the starting point $\vec{r}$.[1]

With the ansatz 3.4 the time dependence of the operators is governed by the time dependence of their prefactors. The time dependence of the prefactors is determined by the use of the Heisenberg equation of motion [79]

$$\partial_t \hat{A}(\vec{r},t) = i\left[\hat{H}, \hat{A}(\vec{r},t)\right] \quad (3.7)$$

for the time derivative of any operator $\hat{A}$. For a calculation with the iterated equation of motion approach an operator basis $\{A_i\}$ has to be chosen. Then the creation operator $\hat{c}^\dagger_{\vec{r},\uparrow}(t)$ is expressed through the ansatz in Eq. 3.4 by the use of these operators. In this way the time dependence is given by the prefactors $h_i(\delta_1,...,t)$. As new monomials are created due to the commutator, the basis $\{A_i\}$ is expanded with every application of the Heisenberg equation.
Consequently the approach is based on the calculation of commutators. At the beginning $t = 0$ the creation operator fulfills $\hat{c}^\dagger_{0,\uparrow}(0,t) = \hat{c}^\dagger_{0,\uparrow}$ with $h_0(0,0) = 1$. Due to translational symmetry site 0 can denote any site of the lattice.

To explain the effect of the commutation on the operators the Hamiltonian is split up into the non-interacting part $\hat{H}_0$ and the interaction $\hat{H}_{int}$. Commutating the creation operator $\hat{c}^\dagger_{0,\uparrow}$ in the one-dimensional model with the non-interacting part $\hat{H}_0$ leads to a simple shift of the operator

$$[\hat{H}_0, \hat{c}^\dagger_{0,\uparrow}] = -J\hat{c}^\dagger_{1,\uparrow} - J\hat{c}^\dagger_{-1,\uparrow}. \quad (3.8)$$

The additional monomials $\hat{c}^\dagger_{1,\uparrow}$ and $\hat{c}^\dagger_{-1,\uparrow}$ are added to the operator basis with the corresponding prefactors $h_0(1,t)$ and $h_0(-1,t)$. A commutation with the interaction term may additionally create or annihi-

---

[1] A specific monomial with given shifts δ_1, δ_2, and δ_3 can be created in several ways. Thus the determination of the distance is not straightforward. To calculate the distance assumptions concerning the operators contained in the monomial have to be made.

3.2 Iterated Equation of Motion Approach

late particle-hole pairs. Starting with $\hat{c}^\dagger_{0,\uparrow}$ the commutation yields

$$[\hat{H}_{int}, \hat{c}^\dagger_{0,\uparrow}] = U\hat{c}^\dagger_{0,\uparrow}\hat{c}^\dagger_{0,\downarrow}\hat{c}_{0,\downarrow} + \dots \qquad (3.9)$$

inducing a monomial with a particle and a particle-hole term. A monomial like this is included in the product $\left(\hat{\uparrow}^\dagger\hat{\uparrow}^\dagger\hat{\downarrow}^\dagger\right)$ of the ansatz 3.4 with the prefactor $h_1(0,0,0,t)$. Further commutations will lead to monomials with a particle and two particle-hole pairs. Iterating this process then leads to the ansatz 3.4. This iterated commutation leads to an expansion in more and more monomials with a higher and higher number of operators involved.

By the help of the ansatz, differential equations for the prefactors $h(\delta_1, \delta_2, \dots, t)$ can be derived by comparing the right and the left hand side of the equation of motion 3.7. In a next step these differential equations are solved numerically by a Runge-Kutta algorithm. Finally the time dependence of the prefactors is reinserted into the ansatz so that the time dependence of $\hat{c}^\dagger_{\vec{r},\uparrow}(t)$ is given. With the time dependence of $\hat{c}(t)$ and $\hat{c}^\dagger(t)$ the one-particle correlation function $G_\sigma(\vec{r}, t)$ can be calculated.

The momentum distribution is derived from $G_\sigma(\vec{r}, t)$ via Fourier transformation. For the jump $\Delta n(t)$ the Fourier transform of the prefactors $h_0(\vec{k}, t)$ for the one particle terms has to be known [72]. After the quench the jump still occurs at the Fermi surface of the non-interacting Hamiltonian FS. The jump is given by the contributions to $G_\sigma(\vec{r}, t)$ that show the slowest decrease in the limit $|\vec{r}| \to \infty$. A product of one-particle terms obeys $\langle \hat{c}^\dagger_{\vec{r},\sigma} \hat{c}_{\vec{0},\sigma} \rangle \propto \frac{1}{r}$ whereas all other products contain higher exponents, such as $\langle : \hat{c}^\dagger_{\vec{r},\uparrow} \hat{c}^\dagger_{\vec{r},\downarrow} \hat{c}_{\vec{r},\downarrow} : ; : \hat{c}^\dagger_{0,\downarrow} \hat{c}_{0,\downarrow} \hat{c}_{0,\uparrow} : \rangle \propto \frac{1}{r^3}$.

Thus the jump is given by the prefactors of the one-particle terms through

$$\Delta n(t) = |h_0(\vec{k}, t)|^2 \Big|_{FS}. \qquad (3.10)$$

3.2.1. Normal Ordering

As the differential equations for the prefactors are deduced by comparing the two sides of the equations of motion a unique representation for the monomials has to be established to make sure that every contribution to a term is mapped to the correct differential equation. For this purpose normal ordering with respect to the Fermi sea is applied. In the iterated equation of motion approach normal ordering is not essential but it simplifies the calculations. A particular advantage of normal ordering is that it allows to distinguish the influence of one-particle terms from the influence of three-particle terms and so on. In this way one-particle effects, for example, cannot be hidden in a term containing three operators as would be possible without normal ordering. Thus the advantage of normal ordered terms is that it can easily be deduced on which states they have a non-vanishing effect. Without normal ordering a monomial containing three operators can also have an effect on a state with one particle as explained below.

Here the usage and advantages of normal ordering are only briefly recalled, as details can be found in various works [80, 81]. In this thesis normal ordered terms are denoted by colons $:\,:$. A normal ordered operator $:\hat{A}:$ describes the fluctuations of the operator $\hat{A}$ around its mean value. Consequently the expectation value of any normal ordered operator vanishes. Especially the expectation value of the normal ordered Hamiltonian reads

$$\langle :\hat{H}: \rangle = 0. \tag{3.11}$$

To translate a monomial in its normal ordered form Wick's theorem [81] can be used. Applying this theorem the contraction of two operators is needed. The contraction is denoted by brackets $\langle\,\rangle$

$$\langle \hat{a}\hat{b} \rangle = \hat{a}\hat{b} - :\hat{a}\hat{b}: \tag{3.12}$$

and describes the correlations of the operators in the considered

3.2 Iterated Equation of Motion Approach

Hamiltonian. A monomial is normal ordered by

$$\hat{c}_1 \hat{c}_2 \ldots \hat{c}_n =: \hat{c}_1 \hat{c}_2 \ldots \hat{c}_n:$$
$$+ : \langle \hat{c}_1 \hat{c}_2 \rangle \ldots \hat{c}_n : + \text{all combinations containing one contraction}$$
$$+ : \langle \hat{c}_1 \hat{c}_2 \rangle \langle \hat{c}_3 \hat{c}_4 \rangle \ldots \hat{c}_n : + \text{all combinations containing two contractions}$$
$$+ \ldots$$
$$+ \text{ all maximally possible contractions}. \tag{3.13}$$

As the reference state is the non-interacting Fermi sea the expectation values of the contractions have to be evaluated with respect to this state. For general creation and annihilation operators

$$\langle \hat{c}_{\vec{r},\sigma} \hat{c}_{\vec{s},\sigma} \rangle = 0 = \langle \hat{c}^\dagger_{\vec{r},\sigma} \hat{c}^\dagger_{\vec{s},\sigma} \rangle \tag{3.14}$$

can be used to simplify the calculations. In the same manner the expectation value of the contraction of two operators with opposite spin vanishes

$$\langle \hat{c}^\dagger_{\vec{r},\sigma} \hat{c}_{\vec{s},\bar{\sigma}} \rangle = 0. \tag{3.15}$$

As an example the monomial $\hat{c}^\dagger_{\vec{r},\uparrow} \hat{c}^\dagger_{\vec{r},\downarrow} \hat{c}_{\vec{r},\downarrow}$ created in the one-dimensional model is considered. Its normal ordered form is derived by

$$\hat{c}^\dagger_{\vec{r},\uparrow} \hat{c}^\dagger_{\vec{r},\downarrow} \hat{c}_{\vec{r},\downarrow} =: \hat{c}^\dagger_{\vec{r},\uparrow} \hat{c}^\dagger_{\vec{r},\downarrow} \hat{c}_{\vec{r},\downarrow} :$$
$$+ : \hat{c}^\dagger_{\vec{r},\uparrow} : \langle \hat{c}^\dagger_{\vec{r},\downarrow} \hat{c}_{\vec{r},\downarrow} \rangle \tag{3.16}$$

where all other possible contractions vanish according to the rules given above. In this example a one-particle term $\propto : \hat{c}^\dagger_{\vec{r},\uparrow} :$ is implicitly included in the three-particle term $\hat{c}^\dagger_{\vec{r},\uparrow} \hat{c}^\dagger_{\vec{r},\downarrow} \hat{c}_{\vec{r},\downarrow}$. Due to normal ordering such hidden effects are avoided.

The contractions are calculated by the use of the Fermi wave vectors via Fourier transform (see below). Having calculated the contraction $\langle \hat{c}^\dagger_{\vec{r},\sigma} \hat{c}_{\vec{s},\sigma} \rangle$ the other contractions can be reduced to this one

Model and Methods

via

$$\langle \hat{c}_{\vec{r},\sigma} \hat{c}^\dagger_{\vec{r},\sigma} \rangle = 1 - \langle \hat{c}^\dagger_{\vec{r},\sigma} \hat{c}_{\vec{r},\sigma} \rangle \qquad (3.17a)$$

$$\langle \hat{c}_{\vec{r},\sigma} \hat{c}^\dagger_{\vec{s},\sigma} \rangle = -\langle \hat{c}^\dagger_{\vec{s},\sigma} \hat{c}_{\vec{r},\sigma} \rangle. \qquad (3.17b)$$

In the following, the contractions for the one-dimensional model are calculated. For the contractions in the two-dimensional model the reader is referred to Sect. 6. For the expectation value of the density term the integral over all occupied states has to be calculated

$$\langle \hat{c}^\dagger_{j,\sigma} \hat{c}_{j,\sigma} \rangle = \frac{1}{2\pi} \int_{-\pi}^{\pi} \Theta(\epsilon_F - \epsilon_k) \, dk \qquad (3.18a)$$

$$= \frac{1}{2\pi} 2 \int_0^{k_F} dk \qquad (3.18b)$$

$$= \frac{1}{\pi} k_F \qquad (3.18c)$$

with the Fermi vector k_F. The expectation value of the hopping term reads

$$\langle \hat{c}^\dagger_{j,\sigma} \hat{c}_{0,\sigma} \rangle = \frac{1}{2\pi} \int_{-\pi}^{\pi} \cos(k \cdot j) \Theta(\epsilon_F - \epsilon_k) \, dk \qquad (3.19)$$

where the sine term of the Fourier transform vanishes due to inversion symmetry. Further simplifications yield

$$\langle \hat{c}^\dagger_{j,\sigma} \hat{c}_{0,\sigma} \rangle = \frac{1}{2\pi} \int_{-k_F}^{k_F} \cos(k \cdot j) \, dk \qquad (3.20a)$$

$$= \frac{1}{\pi} \frac{\sin(k_F \cdot j)}{j}. \qquad (3.20b)$$

The results obtained by this approach pertain directly to the infinite lattice as translational invariance is applied. Thus no finite size effects occur. The approach is based on operators in contrast to related approaches based on recursively constructed Hilbert

3.2 Iterated Equation of Motion Approach

spaces [82]. This simplifies the calculations because the expectation values $\langle 0|\hat{c}(\vec{r},t)\hat{c}^\dagger(0,t)|0\rangle$ are evaluated only once at the specific time instant of interest.

As commutations are used within this approach a linked cluster property is applied. Operators acting on disjoint clusters commute, so that these do not contribute. Thus due to the light cone effect an operator based approach like this has to deal with a finite number of monomials for any finite time t although treating an infinite system in the thermodynamic limit. In contrast to this a method focussing on the time dependence of the states instead of the ones for the operators would have to consider the states in an infinite system, which would refer to an infinite number of terms.

Even though only a finite number of monomials has to be considered this number grows exponentially with the number of commutations. As explained before each commutation creates more and more monomials with more and more operators involved. The number of monomials increases exponentially with a factor of about 3 in the one-dimensional case. Due to the proliferating number of new monomials only a finite number of commutations can be performed. For the one-dimensional model up to 11 commutations are possible with about $5 \cdot 10^5$ monomials and a set of differential equations with $2 \cdot 10^7$ terms on the right hand side.

The monomials occurring for the **first time** in the last commutation lead to an overestimation of the weight loss in the one-particle terms (see Sect. 3.4). Consequently, these are omitted in order to improve the convergence. A calculation performed in this way with m commutations is called an m-**loop** calculation.

3.2.2. Convergence

The differential equations are solved numerically with the initial condition $\hat{c}^\dagger_{\vec{r},\uparrow}(t=0) = 1 \cdot \hat{c}^\dagger_{\vec{r},\uparrow}$ which translates to the initial conditions for the prefactors $h_0(0, t=0) = 1$ and $h_j(\vec{\delta_1}, \vec{\delta_2}, ..., t=0) = 0$ for all other prefactors. Each commutation increments the depth of the hierarchy of the differential equations by one, thus describing

Model and Methods

one order in time t more than the one before. Consequently a calculation with m commutations provides results for $\hat{c}^\dagger(t)$ which are exact up to order t^m. At this point the time t has to be understood as multiplied by an energy scale E^m to obtain a dimensionless quantity. One possible energy scale would be the maximum $E = \max(U, W)$. However, the equations are not solved by series expansion. Instead the set of differential equations is solved numerically. In this way also higher orders in the time t are generated (see Sect. 3.5).

For the one-dimensional model with 11 loops the results are converged up to $t \approx 10/W$. To quantify the time t up to which the results of the corresponding calculation are converged, calculations with different numbers of loops are performed and their results are compared. The precise value depends on the details of the calculation like the interaction strength and the filling factor. In the following the result is regarded as converged if the differences between calculations with different numbers of loops are smaller than the thickness of the line.

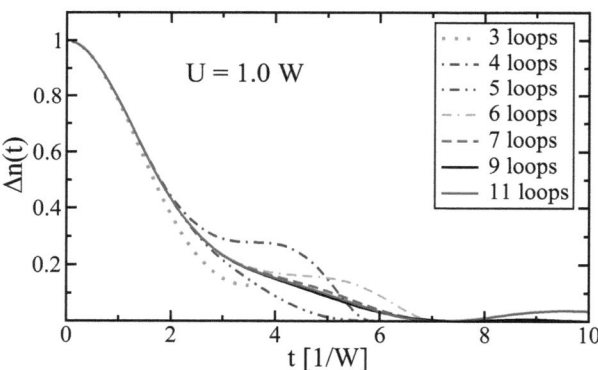

Fig. 3.2.: Jump $\Delta n(t)$ for the one-dimensional model and an interaction strength $U = 1.0W$ in dependence on the time t. The results are shown for calculations with different loop numbers. The results of different calculations are compared to each other to quantify the convergence. Each loop increases the range of convergence.

3.2 Iterated Equation of Motion Approach

As exemplary result the jump $\Delta n(t)$ for a quench to $U = 1.0W$ is depicted in Fig. 3.2. For $t = 0$ the jump starts at $\Delta n(0) = 1$ as it should be according to the jump in the non-interacting model. The jump is shown for various numbers of loops from three loops to 11 loops. At the beginning all curves overlap. Then the curves deviate at different instants of the time t. Comparing the 3-loop calculation with the 4-loop calculation it can be seen that the two curves coincide up to $t \approx 1.2/W$. Thus the results of these calculations can be considered as converged up to this time t. In comparison with the 5-loop calculation the 4-loop calculation is exact up to $t \approx 2/W$. In this way the range of convergence is increased with increasing loop number until the maximal loop number 11 is reached with a range of convergence of about $t \approx 7.4/W$ for this rather large value of U. As additional test for the convergence the expectation value $\langle 0|\hat{n}_0(t)|0\rangle$ for finding a particle at site 0 can be calculated. This value should be constant and equal to the filling factor as no particles are inserted or removed within the calculation. The local expectation value is calculated by the product of all monomials in the corresponding operator basis

$$\langle 0|\hat{n}_0(t)|0\rangle = \langle 0|\hat{c}^\dagger_{0,\uparrow}(t)\hat{c}_{0,\uparrow}(t)|0\rangle \tag{3.21}$$

$$= \langle 0|\left(h_0(0,t)\hat{c}^\dagger_{0,\uparrow} + h_0(1,t)\hat{c}^\dagger_{1,\uparrow} + h_1(0,0,0,t):\hat{c}^\dagger_{0,\uparrow}\hat{c}^\dagger_{0,\downarrow}\hat{c}_{0,\downarrow}: + ...\right) \tag{3.22}$$

$$\left(h^*_0(0,t)\hat{c}_{0,\uparrow} + h^*_0(1,t)\hat{c}_{1,\uparrow} + h^*_1(0,0,0,t):\hat{c}^\dagger_{0,\downarrow}\hat{c}_{0,\downarrow}\hat{c}_{0,\uparrow}: + ...\right)|0\rangle.$$

Corresponding results are shown in Fig. 3.3. As can be seen the results deviate from their initial value for larger times. Due to the finite number of loops the iterated equation of motion approach does not necessarily obey unitarity, so that these deviations are to be attributed to the breakdown of the unitarity (see Sect. 4.1).

Exemplary results showing the loss of unitarity can be found in Fig. 3.3. In this figure the expectation value is shown for an interaction quench to $U = 1.5W$ and various numbers of loops. As the evaluation of the expectation value involves the product of all terms, the evaluation of the expectation value is more demanding than the one for the jump itself. Thus the maximal loop number for

Model and Methods

which such a calculation is possible is reduced to nine loops. For the sake of clarity the results are shown from five to nine loops. Clearly all curves start with $\langle \hat{n}_0(0) \rangle = 0.5$. As there are neither particles inserted nor removed in the calculation and translational invariance is preserved this value should stay constant to 0.5 for the half-filled model. For longer times all curves deviate from this value. With increasing numbers of loops the point where the curves deviate from their initial value is shifted to longer times. The 5-loop calculation becomes unreliable at $t \approx 1.25/W$, whereas the 9-loop calculation is converged up to $t \approx 3/W$ for this interaction strength.

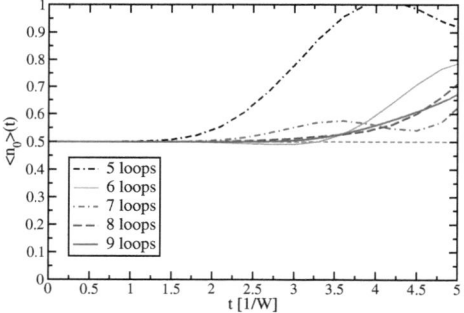

Fig. 3.3.: Expectation value $\langle \hat{n}_0(t) \rangle$ for an interaction strength $U = 1.5W$ in dependence on the time t. The different curves represent results for various numbers of loops. For half-filling the expectation value should be equal to 0.5 (dashed brown line). With the number of loops the time up to which the result equals its initial value is increased. Deviations from this value are to be attributed to non-unitarity effects.

Fig. 3.4.: Local expectation value $\langle \hat{n}_0 \rangle(t)$ as derived in a 9-loop calculation for the half-filled model. The results are shown for various interaction strengths U.

As the range of convergence is strongly depending on the strength of the quench, it has to be checked for every value of U separately. A comparison of the ranges of convergence for various interaction strengths U is depicted in Fig. 3.4. In this figure the drastic decrease in the range of convergence on increasing U can

3.2 Iterated Equation of Motion Approach

be seen. The results are derived in a 9-loop calculation. The expectation value for a quench to $U = 2.0W$ deviates already for $t \approx 3/W$ from its initial value, whereas the results for $U = 0.5W$ are exact up to $t \approx 4.5/W$. For smaller $U \approx 0.1W$ the expectation value is constant up to $t = 8/W$ and beyond.
As explained the results for the jump are much more robust than the ones for the expectation value. This may be attributed to the fact, that in contrast to the evaluation of the expectation value, the calculation of the jump includes only terms at the Fermi surface. The deviations of the expecation value from the filling factor are the most rigorous bounds for the convergence.

As nearly no assumptions are made concerning the model under study, the presented approach is very flexible and can be applied to various models. The method allows to study other lattices and other observables as long as these can be expressed in terms of creation and annihilation operators. Besides, other initial states like mixed states can be used, so that the influence of different temperatures on the relaxation can be investigated. In contrast to perturbative approaches the method presented in this thesis is capable of arbitrarily large interaction strengths U.
The only restriction posed on the approach is that it can only be applied to quenches where the interaction and the hopping can be assumed to be short-ranged processes. Additionally the description is limited to short and intermediate times after the quench due to computational efforts.
In the range where the results are converged the method yields exact results.

3.3. Implementation

Due to the exponentially growing number of terms resulting in a vast amount of commutations the calculations are performed by a computer. As the requirements concerning computational time and memory for setting up the differential equations are totally different from the ones for solving them, the two tasks are performed by distinct programmes. Both programmes are implemented in C++.

The monomials appearing during the calculations are represented by objects of the class **term**.
This class contains a vector consisting of the operators forming the monomial and a complex prefactor. Member functions coordinate tasks such as the creation of terms, manipulation of them and comparing two terms. The operators themselves are encoded in the class **operator**. This class contains information concerning the spin, the coordinates and the type of a single operator. Additionally a flag is used to simplify the normal ordering. The product of two normal ordered terms is translated into its normal ordered form by considering only pairings between one operator of the first term with another one from the second term [83, 84]. Pairings between operators of the same term vanish as these are already considered in the normal ordering of the term itself.
The contributions to the differential equations of the monomials are kept in instants of the datatype **struct** containing the prefactor and an identifier which encodes the corresponding monomial for the right hand side of the differential equation. The general structure of the program is depicted in Fig. 3.5.
In a first step, the Hamiltonian and the starting operator $\hat{c}^\dagger_{0,\uparrow}$ are initialized.

In the program the monomials considered in the corresponding calculation are kept in a vector named **listterms**. This list contains the monomials which have already been commuted. An additional **agenda list** manages the monomials which are to be commuted within the current loop.

3.3 Implementation

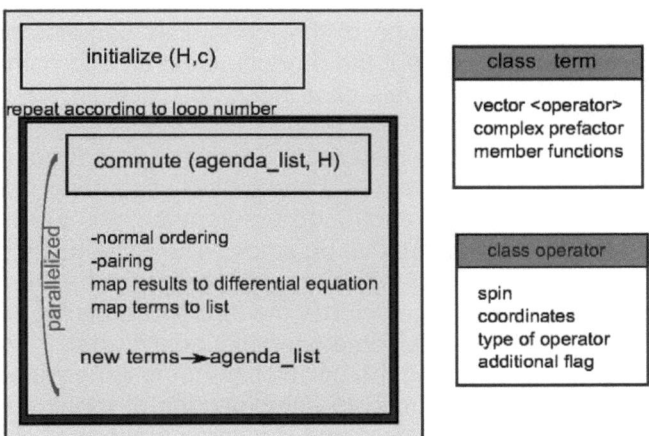

Fig. 3.5.: General structure of the program implemented for the iterated equation of motion approach. First the Hamiltonian and the vector containing the monomials are initialized. Afterwards the monomials are commuted. This step is performed in parallel and is repeated according to the maximal loop number.

In each loop the monomials of the corresponding agenda list are commuted with the Hamiltonian. For this part the following formula is used simplifying the commutation of two monomials with arbitrary length. Two monomials $\hat{A}$ and $\hat{B}$ consisting of fermionic operators $\hat{a}_i$ and $\hat{b}_j$ yield

$$[\hat{A},\hat{B}] = \left[\prod_{i=1}^{n} \hat{a}_i, \prod_{j=1}^{m} \hat{b}_j\right] \tag{3.23a}$$

$$= \sum_{k=1}^{n}\sum_{\ell=1}^{m} (-1)^{m(n-k)}(-1)^{\ell-1} \left(\prod_{i=1}^{k-1} \hat{a}_i \prod_{j=1}^{\ell-1} \hat{b}_j \{\hat{a}_k,\hat{b}_\ell\} \prod_{r=\ell+1}^{m} \hat{b}_r \prod_{s=k+1}^{n} \hat{a}_s \right) \tag{3.23b}$$

with the anticommutator denoted by braces. This formula works for all cases where at least one of the monomials contains an even number of operators. Only if the product mn is an odd number the formula breaks down but this case does not occur in the present thesis.

After the commutation the resulting monomials are normal ordered by the use of Wick's theorem [81] and multiply occurring terms are combined. At the end of each loop the contributions are mapped to the corresponding differential equation. The monomials created during the loop are added to the agenda list for the next loop. For the sake of computational efficiency the loops for different monomials are performed in parallel on a computational cluster.

Due to the normal ordering the calculation of the expectation value $\langle \hat{c}(\vec{r},t)\hat{c}^\dagger(0,t)\rangle$ can be reduced to combinations of monomials with the same number of operators and the same number of operators acting on a state with the spin pointing upwards. In all other cases there is always a vanishing contraction.

To keep the calculations as efficient as possible the monomials are stored in a structure taking care of these properties by storing them in a vector where the entries are grouped according to their number of operators. Within each group an additional vector structure keeps track of the number of operators with a spin pointing upwards. In this way only products of two terms within each subgroup have to be evaluated, which simplifies the calculation significantly.

Finally, the differential equations are solved by a fourth order Runge-Kutta algorithm. Having calculated the time dependent operators $\hat{c}^\dagger$ and $\hat{c}$, the jump, local expectation value and the whole momentum distribution can be determined. The momentum distribution is derived by numerical Fourier transformation of the one-particle correlation function (3.3). One advantage of the approach lies in the fact that the observables, such as the local expectation value, have to be computed only for the time instants of interest. If the value of $\langle \hat{n}_0 \rangle$ at $t = 5/W$ has to be known, the differential equations can easily be solved for $t \leq 5/W$ and then the expectation value can be calculated only for $t = 5/W$.

3.4. Self-Consistent Truncation

To illustrate how the convergence is improved by neglecting the monomials appearing in the last commutation for the first time the time evolution of the non-interacting model ($U = 0$) is considered. For simplicity the results for a calculation with three commutations is studied. At the beginning ($t = 0$) the creation operator reads $\hat{c}^\dagger_{0,\uparrow}(t) = \hat{c}^\dagger_{0,\uparrow}$. Commuting $\hat{c}^\dagger_{0,\uparrow}$ with the non-interacting Hamiltonian $\hat{H}_0$ leads to two additional monomials $\hat{c}^\dagger_{-1,\uparrow}$ and $\hat{c}^\dagger_{1,\uparrow}$ through

$$[\hat{H}_0, \hat{c}^\dagger_{0,\uparrow}] = -J\hat{c}^\dagger_{-1,\uparrow} - J\hat{c}^\dagger_{1,\uparrow}. \tag{3.24}$$

In the next step these new operators have to be commuted once more

$$[\hat{H}_0, \hat{c}^\dagger_{1,\uparrow}] = -J\hat{c}^\dagger_{2,\uparrow} - J\hat{c}^\dagger_{0,\uparrow} \tag{3.25a}$$

$$[\hat{H}_0, \hat{c}^\dagger_{-1,\uparrow}] = -J\hat{c}^\dagger_{-2,\uparrow} - J\hat{c}^\dagger_{0,\uparrow} \tag{3.25b}$$

with the newly created operators $\hat{c}^\dagger_{2,\uparrow}$ and $\hat{c}^\dagger_{-2,\uparrow}$. Then the operators created in the second commutation are considered. But as this is the last commutation, all monomials created in this iteration are neglected, leading to

$$[\hat{H}_0, \hat{c}^\dagger_{2,\uparrow}] = -J\cancel{\hat{c}^\dagger_{3,\uparrow}} - J\hat{c}^\dagger_{1,\uparrow} \tag{3.26a}$$

$$\text{and}\, [\hat{H}_0, \hat{c}^\dagger_{-2,\uparrow}] = -J\cancel{\hat{c}^\dagger_{-3,\uparrow}} - J\hat{c}^\dagger_{-1,\uparrow}. \tag{3.26b}$$

Now that the operators describing the time dependence in a 3-loop calculation are determined, the time dependence of the creation operator can formally be written down in the ansatz

$$\begin{aligned}\hat{c}^\dagger_{0,\uparrow}(t) =& h_0(0,t)\hat{c}^\dagger_{0,\uparrow} + h_0(1,t)\hat{c}^\dagger_{1,\uparrow} + h_0(-1,t)\hat{c}^\dagger_{-1,\uparrow} \\ & + h_0(2,t)\hat{c}^\dagger_{2,\uparrow} + h_0(-2,t)\hat{c}^\dagger_{-2,\uparrow}\end{aligned} \tag{3.27}$$

Model and Methods

in which the time dependent prefactors are given by comparing both sides of the equation of motion. The corresponding differential equations read

$$\partial_t h_0(0,t) = -Jih_0(1,t) - Jih_0(-1,t)$$
$$\partial_t h_0(1,t) = -Jih_0(0,t) - Jih_0(2,t)$$
$$\partial_t h_0(-1,t) = -Jih_0(0,t) - Jih_0(-2,t)$$
$$\partial_t h_0(2,t) = -Jih_0(1,t)$$
$$\partial_t h_0(-2,t) = -Jih_0(-1,t) \tag{3.28}$$

with the initial conditions $h_0(0,0) = 1$ and $h_0(r,0) = 0 \,\forall r \neq 0$. For the non-interacting model the jump is constant $\Delta n(t) = 1$ and the expectation value reads

$$\langle \hat{n}_0(t) \rangle = 0.5 = \text{const.} \tag{3.29}$$

As explained in Sect. 3.2.2 a 3-loop calculation is exact up to third order in time t. For clarity the prefactors are expanded in time in a power series up to $O(t^4)$, which yields

$$h_0(0,t) = 1 - J^2 t^2 + O(t^4) \tag{3.30a}$$
$$h_0(1,t) = -Jit + \frac{1}{2}J^3 it^3 + O(t^4) = h_0(-1,t) \tag{3.30b}$$
$$h_0(2,t) = -\frac{1}{2}J^2 t^2 + O(t^4) = h_0(-2,t) \tag{3.30c}$$

in which the prefactors are either imaginary or real numbers. In this expansion the jump is given by

$$\Delta n(t) = |h_0(0,t) - 2h_0(2,t)| = 1 + O(t^4) \tag{3.31}$$

and the expectation value yields

$$\langle \hat{n}_0(t) \rangle = 0.5 + O(t^4) \tag{3.32}$$

3.4 Self-Consistent Truncation

as it should be. In the presented approach the differential equations are not solved by such an expansion but numerically by a Runge-Kutta algorithm, so that also higher orders in t are created.

Such a calculation where in the last commutation only monomials which appeared already in the preceding iteration are considered can be seen as a self-consistent truncation scheme. In the following a calculation based on this truncation scheme with m iterations is referred to as m-loop calculation. This truncation scheme is contrasted to the full calculation where all appearing operators, i.e., also $h_0(3,t)\hat{c}^\dagger_{3,\uparrow}$ and $h_0(-3,t)\hat{c}^\dagger_{-3,\uparrow}$ in the calculation with three commutations, are taken into account. This leads to additional terms for the differential equations, which read

$$\partial_t h_0(3,t) = -Ji h_0(2,t) \qquad (3.33)$$
$$\partial_t h_0(-3,t) = -Ji h_0(-2,t). \qquad (3.34)$$

As this calculation also includes three commutations the result is still exact up to third order t^3. But in higher orders, as produced by the use of the Runge-Kutta algorithm, the results differ from the ones of the self-consistent approach. The effect of higher orders can be observed in a comparison of the two calculations with three commutations as shown in Fig. 3.6. In this figure the jump $\Delta n(t)$ and the local expectation value are depicted as function of the time t. As can be seen, the jump $\Delta n(t)$ (upper curve) stays constant at unity as expected for the non-interacting model. The expectation value (lower curve) starts at the desired value of 0.5 for both curves. The result of the full calculation deviates from this value for $t \approx 2.0/W$ whereas the value in the self-consistent calculation is still constant. Consequently the self-consistent approach describes the evolution much better than the full calculation considering also the new appearing terms of the last commutation. This result is exemplary for a self-consistent calculation with an odd number of loops.

Model and Methods

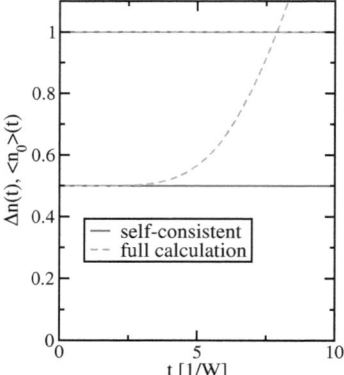 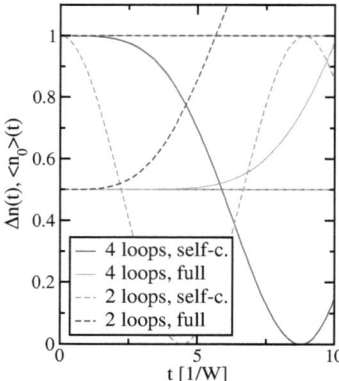

Fig. 3.6.: Results for the jump (upper curves) and the expectation value $\langle \hat{n}_0(t) \rangle$ (lower curves) for a calculation with three loops opposed to a full calculation with three commutations considering all terms. For the non-interacting model the jump and the expectation value should be constant. The results are exemplary for calculations with odd numbers of commutations.

Fig. 3.7.: Comparison of the self-consistent loop calculation with the full calculation for even numbers of commutations. For the jump (upper curves) the results of the full calculations stay constant at the desired value of 1.0 whereas the loop calculations deviate. In contrast to this the expectation value (lower curves) is better described by the self-consistent loop calculations.

The same comparison for even numbers of commutations can be found in Fig. 3.7. On this level the jump stays constant for the full calculations whereas the jump derived in the self-consistent loop calculations deviates. But concerning the expectation value $\langle \hat{n}_0(t) \rangle$ the result for the self-consistent calculations are much better than the ones for the full calculations. The self-consistent calculations lead to a constant expectation value as desired.

In conclusion, the self-consistent loop calculation yields better results for the expectation value in both cases, for even and odd numbers of commutations. Comparing the results for an odd number of

3.4 Self-Consistent Truncation

commutations depicted in Fig. 3.6 to the results for an even number of commutations given in Fig. 3.7, it can be concluded that the jump $\Delta n(t)$ is best described by the loop calculation if an odd number of commutations is applied. In the case of an odd number of loops the jump calculated for the non-interacting case stays constant to 1.0 as it should be.

So far the non-interacting model with $U = 0$ has been discussed. In the following a quench to a nominal interaction $U = 1.0W$ is discussed within the self-consistent loop approach and the full calculation. The expectation value for these calculations with different numbers of commutations is shown in Fig. 3.8. The computational effort for evaluating the expectation value of a full calculation with $n-1$ commutations is comparable to the one for a self-consistent loop calculation with n loops. Thus the full calculation with three commutations has to be compared to the self-consistent 4-loop calculation and so on.

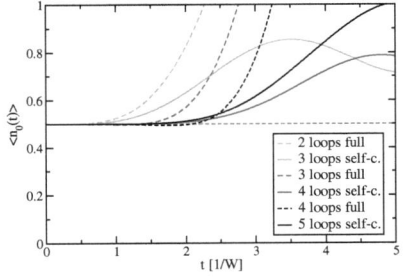

Fig. 3.8.: Local expectation value for a quench to $U = 1.0W$ in dependence on time. The solid lines represent results for the self-consistent loop calculation with various numbers of loops. The dashed lines depict results for the corresponding calculations considering all monomials created during the last commutation.

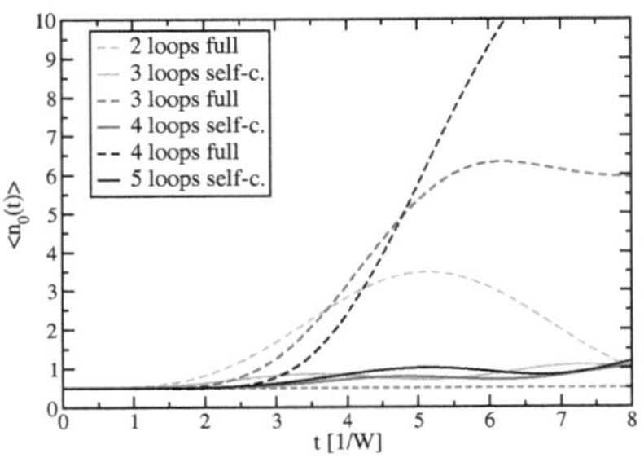

Fig. 3.9.: Results of Fig. 3.8 plotted with a larger range for the y-axis. Once deviating from the initial value the results for the full calculations shoot up drastically. In contrast to this the results for the self-consistent loop calculations stay fairly small.

Comparing the two truncation schemes, it can clearly be seen in Fig. 3.8 that the convergence of the self-consistent calculations is better than the one for the full calculation. The full calculation with three commutations deviating for $t \approx 1.1/W$ has to be compared to the 4-loop calculation which deviates at $t \approx 1.7/W$. Besides, it is clearly visible how each commutation increases the convergence. Apart from the range of convergence the loop approach has another advantage, which can be seen in Fig. 3.9. In this figure the same results are shown with a larger range on the y-axis. The results for the full calculations shoot up to high values very quickly. In contrast to this the results for the self-consistent loop calculations stay much smaller for the observed times.

As the jump is used as a sensitive probe for the dynamics after the quench, results for this observable obtained with the two truncation schemes are compared in Fig. 3.10.

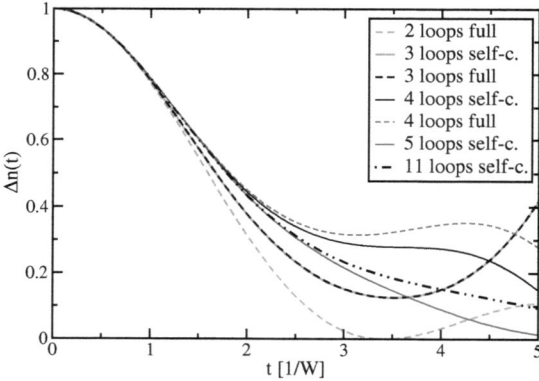

Fig. 3.10.: Jump $\Delta n(t)$ as derived in self-consistent loop calculations opposed to the corresponding full calculations considering all monomials created in the calculations (especially the monomials created during the last commutation). Dashed-dotted line: Result of an 11-loop calculation given as reference.

3.4 Self-Consistent Truncation

The results are shown for calculations with two to five commutations, with the result of the corresponding 11-loop calculation used as reference. For three commutations the curves for the two approaches coincide, as up to this point only a few terms are included on calculating the jump. Increasing the number of commutations to four, the results lie on top of the reference curve up to $t \approx 1.6/W$. The 4-loop calculation is still valid until $t \approx 1.7/W$ and the corresponding 5-loop calculation is reliable up to $t \approx 2.4/W$. Altogether the self-consistent loop calculation shows a better performance also for finite interaction strengths. Consequently this truncation scheme denoted as m-loop calculation is used henceforth in this thesis.

3.5. Runaway Time

In Sect. 3.2.2 it was explained that due to the structure of the differential equations and the initial conditions each commutation comprises one order in time t. To illustrate how each loop improves the results this section provides a discussion of the deviations of calculations with different numbers of loops where the result of the 11-loop calculation is used as reference. The absolute difference of $\Delta n(t)$ obtained in an m-loop calculation from the one gained within 11 loops is shown in a double logarithmic plot in Fig. 3.11. The results are calculated for an interaction strength of $U = 1.0W$ and they display the characteristic behavior of the absolute difference.

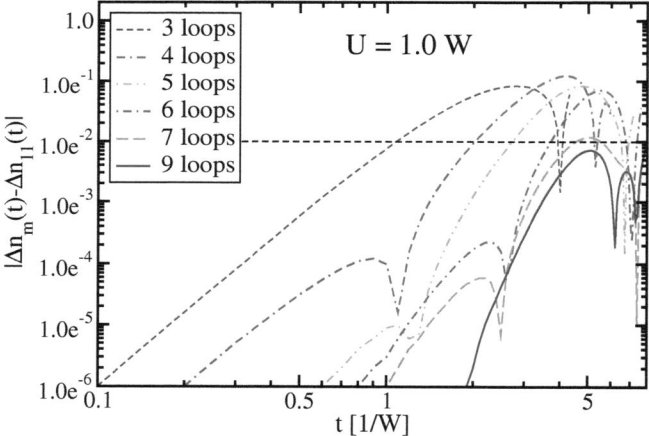

Fig. 3.11.: Absolute difference of an m-loop calculation from the corresponding 11-loop calculation against time t. The curves are calculated for $U = 1.0W$ and different numbers of loops m. The dashed line depicts a value of 0.01 which is used as threshold in the determination of the runaway time t_{runaway}.

3.5 Runaway Time

Obviously the differences show an increase with time t with dips at the times where the curves occasionally coincide. For further discussions a runaway time is defined as the time $t_{runaway}$ beyond which the difference in the results exceeds a certain threshold. In a first attempt the threshold is set to 0.01. This threshold is depicted as dashed line in Fig. 3.11. The runaway times determined in this manner are depicted in Fig. 3.12. The plot shows the inverse runaway time dependent on the inverse loop number $\frac{1}{m}$.

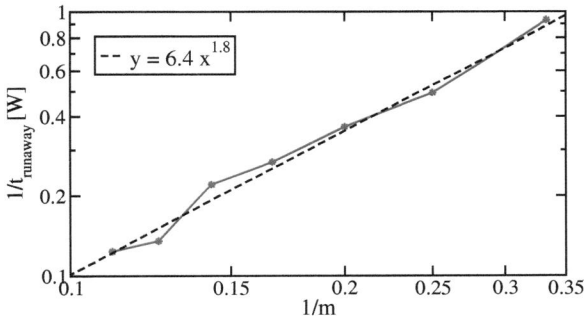

Fig. 3.12.: Inverse runaway time against the inverse number of loops in a double logarithmic plot. The dashed line shows a power law fit to the data with an exponent of about 1.8.

The data shows a power law decrease of the inverse runaway time on increasing m thus indicating a power law increase of the runaway time. The dashed line shows a power law fit to the data with an exponent of about 1.8. Surprisingly the convergence is superlinear which is an advantage for practical use of the method. From the structure of the differential equations a linear correlation would be expected. To understand this large exponent further

Model and Methods

analysis is needed. Of course, the choice of the threshold is arbitrary, but it illustrates the convergence of the results for an increasing loop number. In particular it depicts the behavior in the limit $m \to \infty$. For $m \to \infty$ the inverse runaway time tends to zero, suggesting that the result becomes exact in this limit. To clarify to what extent the runaway time is influenced by the choice of the threshold, the runaway times for two thresholds are depicted in Fig. 3.13. In this figure the inverse runawaytime is shown as function of the inverse loop number in a double logarithmic plot. Although the exact values are determined by the threshold, the runaway time behaves similarly for all thresholds. The curves decrease on increasing m and the behavior in the limit $m \to \infty$ is recovered. Although the exponents differ slightly, a superlinear convergence can be deduced from all fits. If a threshold of 0.05 is chosen, the exponent takes a value of about 1.87.

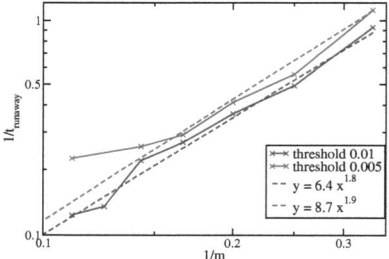

Fig. 3.13.: Inverse runaway times for different values of the threshold. The corresponding fits are given as dashed curves with exponents of 1.8 (blue curve) and 1.9 (red curve).

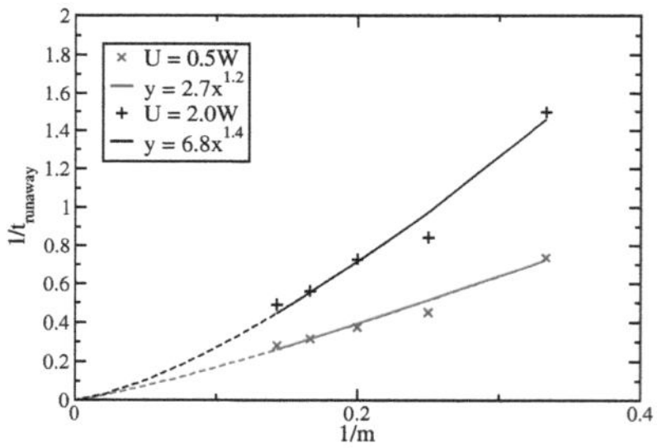

Fig. 3.14.: Inverse runaway time for the local expectation value $\langle 0|\hat{c}^\dagger_{0,\uparrow}\hat{c}_{0,\uparrow}|0\rangle$ in a linear plot. The runaway times are shown for a threshold of 0.01 and various values of the interaction U. The fits indicate convergence towards zero for an infinite number of loops and a superlinear convergence.

As seen before the convergence of the local expectation value $\langle 0|\hat{c}^\dagger_{0,\uparrow}\hat{c}_{0,\uparrow}|0\rangle(t)$ for finding a particle at site 0 is worse than the

3.5 Runaway Time

one for the jump $\Delta n(t)$. The corresponding runaway times for a threshold of 0.01 and various values of U are displayed in Fig. 3.14 in a linear plot. Power law fits to the data indicate exponents of about 1.2 for an interaction of $U = 0.5W$ and an exponent of about 1.4 for the rather strong interaction of $U = 2.0W$. The fits indicate that the inverse runawaytime vanishes for an infinite number of loops. Although the convergence is weaker than for the jump - as expected from Sect. 3.1.1 - the runaway time is still superlinear.

Model and Methods

4. Variants of the Approach

In this section the non-unitarity appearing during the calculations is discussed further. Furthermore, variants of the iterated equation of motion approach developed to circumvent this problem are presented. As none of these approaches yielded satisfying improvements concerning unitarity, these approaches are not pursued further in this thesis. However, these variants are presented in the following as these could be the basis for further studies adressing issues different from the ones discussed in this thesis. Besides, additional truncation schemes developed for the iterated equation of motion approach are explained in App. B.

4.1. Matrix Approaches

For a 2-loop calculation the differential equations can in general be written in matrix form

$$\partial_t \vec{v} = i\underline{M}\vec{v} \qquad (4.1)$$

with the matrix $\underline{M}$ and the vector $\vec{v}$ containing the prefactors for the terms $h_0(0,t) : \hat{c}^\dagger_{0,\uparrow} :$, $h_0(-1,t) : \hat{c}^\dagger_{-1,\uparrow} :$, $h_0(1,t) : \hat{c}^\dagger_{1,\uparrow} :$ and $h_1(0,0,0,t) : \hat{c}^\dagger_{0,\uparrow}\hat{c}^\dagger_{0,\downarrow}\hat{c}_{0,\downarrow} :$ from top to bottom.

4.1.1. Non-unitarity

As an example the half-filled case is considered where the time dependence is given by the matrix

$$\underline{M} = \begin{pmatrix} 0 & J & J & \frac{U}{4} \\ J & 0 & 0 & 0 \\ J & 0 & 0 & 0 \\ U & 0 & 0 & 0 \end{pmatrix}. \qquad (4.2)$$

Having achieved a matrix form the problem can be translated to an eigenvalue problem $\lambda \vec{v} = \underline{M}\vec{v}$ with the eigenvalues λ_i. For the 2-loop calculation these are the the two-fold degenerate eigenvalue 0 and the eigenvalues $\lambda = \pm \sqrt{\frac{U^2}{4} + 2J^2}$. Surprisingly calculations with an increased number of commutations reveal imaginary eigenvalues. For half-filling the eigenvalues appear in pairs $\pm \lambda$ due to particle-hole symmetry. This property is lost upon doping. Based on the eigenvectors v_i and the eigenvalues λ_i the general structure of the time dependent operator $\hat{c}^\dagger_{0,\uparrow}(t)$ reads

$$v(t) = \alpha_1 v_1 e^{i\lambda_1 t} + ... + \alpha_n v_n e^{i\lambda_n t} \qquad (4.3)$$

where the prefactors α_i ensure the initial conditions. Imaginary eigenvalues λ_i lead to exponentially increasing or decreasing terms. They definitely indicate the breakdown of the unitarity. Thus the rapidly increasing curves for the jump and the expectation value are partly due to the imaginary eigenvalues.

4.1.2. Different Scalar Products

Up to now the equations for the prefactors are derived by comparing both sides of the Heisenberg equations of motion. A different approach to determine the prefactors is to set up an operator basis and represent the prefactors within this basis. The operators forming the basis are assigned with a time dependent prefactor h. In this

4.1 Matrix Approaches

way the prefactors can be written as entries of a vector $\vec{v}$.

$$\vec{v} = \begin{pmatrix} h_0(0,t) \\ h_0(1,t) \\ \vdots \end{pmatrix} \quad (4.4)$$

With the use of the Liouville superoperator $\mathcal{L}\hat{A} = [\hat{H},\hat{A}]$ equations for the operators can be set up. These are then transformed into a matrix equation for the vector $\vec{v}$ by multiplying this equation with the vector of operators. Thus a matrix $\underline{M}$ describing the Liouville superoperator is obtained. Additionally the norm matrix $\underline{N}$ is introduced. A generalized eigenvalue problem

$$\underline{N}\partial_t \vec{a} = i\underline{M}\vec{a} \quad (4.5)$$

is achieved. Due to the representation in an operator basis the choice of the scalar product is essential.
The use of a different scalar product (.|.) within the matrix representation may lead to improved ranges of convergence. The effect of different choices for the scalar product is discussed in this section.

In contrast to conventional Liouville approaches, the operators used in the approach presented here are not created in a Krylov-basis. Instead a set of operators to be considered is chosen before the calculation. But the set of operators is chosen such that an m-loop calculation stays inside the set of operators. The corresponding prefactors are given as the entries of the vector v_i as explained above. With these prefactors the matrices are defined as

$$N_{ij} = (v_i|v_j) \quad (4.6a)$$
$$M_{ij} = (v_i|\mathcal{L}v_j) \quad (4.6b)$$

with the scalar product of choice (.|.).
A possible choice for the scalar product of two operators $\hat{A}$ and $\hat{B}$ is

$$(\hat{A}|\hat{B}) = \langle 0|\hat{A}^\dagger \hat{B}|0\rangle. \quad (4.7)$$

Variants of the Approach

Taking the expectation value with respect to the Fermi sea leads to a rather complicated form for the norm matrix N

$$N = \begin{pmatrix} \frac{1}{2} & -\frac{1}{\pi} & -\frac{1}{\pi} & 0 \\ -\frac{1}{\pi} & \frac{1}{2} & 0 & 0 \\ -\frac{1}{\pi} & 0 & \frac{1}{2} & 0 \\ 0 & 0 & 0 & \frac{1}{8} \end{pmatrix} \qquad (4.8)$$

describing a system with the four operators appearing in the 2-loop calculation. As the scalar product is determined by the expectation value the curves derived within this approach are labelled by *expt* in the plots. In this description the matrix $\underline{M}$ takes a more complicated form. Many more non-vanishing prefactors α_i appear so that more monomials have to be considered leading to a much more demanding calculation. To improve the results one more iteration is performed. On this level 10 additional operators

$$\hat{c}^\dagger_{-2,\uparrow}, \quad \hat{c}^\dagger_{2,\uparrow}$$
$$:\hat{c}^\dagger_{-1,\uparrow}\hat{c}^\dagger_{-1,\downarrow}\hat{c}_{-1,\downarrow}:, \quad :\hat{c}^\dagger_{1,\uparrow}\hat{c}^\dagger_{1,\downarrow}\hat{c}_{1,\downarrow}:$$
$$:\hat{c}^\dagger_{-1,\uparrow}\hat{c}^\dagger_{0,\downarrow}\hat{c}_{0,\downarrow}:, \quad :\hat{c}^\dagger_{0,\uparrow}\hat{c}^\dagger_{-1,\downarrow}\hat{c}_{0,\downarrow}:, \quad :\hat{c}^\dagger_{0,\uparrow}\hat{c}^\dagger_{0,\downarrow}\hat{c}_{-1,\downarrow}:$$
$$:\hat{c}^\dagger_{1,\uparrow}\hat{c}^\dagger_{0,\downarrow}\hat{c}_{0,\downarrow}:, \quad :\hat{c}^\dagger_{0,\uparrow}\hat{c}^\dagger_{1,\downarrow}\hat{c}_{0,\downarrow}:, \quad :\hat{c}^\dagger_{0,\uparrow}\hat{c}^\dagger_{0,\downarrow}\hat{c}_{1,\downarrow}:$$

appear leading to a 14×14 matrix. Already on this level complex eigenvalues occur. In this representation the Liouville superoperator is not hermitian with respect to Eq. 4.7. Thus this calculation also faces non-unitarity effects.

Results for the jump and the expectation value derived within this calculation are given in Fig. 4.1. These results are opposed to results obtained in a 3-loop calculation which is labelled as 'direct' approach. The result for the jump deviates from the reference curve much earlier (at about $t = 1.6/W$) than the one of the 3-loop calculation, which is reliable up to $t \approx 2.8/W$. The expectation value also deviates much earlier (at around $t \approx 1/W$) than in the 3-loop calculation. Thus the convergence is not improved by this approach.

4.1 Matrix Approaches

4.1.2.1. Scalar Product on Basis of the Anticommutator

Another choice for the scalar product makes use of the anticommutator of the operators by

$$(\hat{A}|\hat{B}) = \text{Tr}\left(\{\hat{A}^\dagger, \hat{B}\}\right)\frac{1}{\text{Tr}\mathbb{1}} \qquad (4.9)$$

with the trace Tr. Following this definition the use of the anticommutator

$$\{\hat{A}^\dagger, [\hat{H}, \hat{B}]\} = \hat{A}^\dagger(\hat{H}\hat{B} - \hat{B}\hat{H}) + (\hat{H}\hat{B} - \hat{B}\hat{H})\hat{A}^\dagger \quad \text{and} \qquad (4.10\text{a})$$
$$\{[\hat{H}, \hat{A}]^\dagger, \hat{B}\} = (-\hat{H}\hat{A}^\dagger + \hat{A}^\dagger\hat{H})\hat{B} + \hat{B}(\hat{A}^\dagger\hat{H} - \hat{H}\hat{A}^\dagger) \qquad (4.10\text{b})$$

yields

$$\left(\hat{A}|[\hat{H}, \hat{B}]\right) = \left([\hat{H}, \hat{A}]|\hat{B}\right) \qquad (4.11)$$

as the trace allows cyclic changes of the operators. Thus on the operator level the use of this scalar product should yield a unitarity preserving description.

However, the results reveal non-unitary behavior (see Fig. 4.2). This is due to the difference in the definition of unitarity on the operator level and in the basis of the actual states. Even though the matrix is unitary in terms of the operators this does not have to be true on the basis of the states.

Formulated in the basis of the states unitary transformations

$$\hat{c}_0(t) = U^\dagger \hat{c}_0(0) U \qquad (4.12)$$

are characterized by $\hat{c}_0(t)\hat{c}_0^\dagger(t) + \hat{c}_0^\dagger(t)\hat{c}_0(t) = U^\dagger\{\hat{c}_0, \hat{c}_0^\dagger\}U = U^\dagger U = 1$ in contrast to the results of the matrix approaches.

Comparing the results for the expectation value derived within this approach (labelled *anticomm*) with the *direct* method, the ranges of convergence are comparable (see Fig. 4.2). However, the calculations based on the scalar product with the anticommutator are much more demanding concerning computational time. As the re-

Variants of the Approach

sults are not significantly improved by this approach and the non-unitarity is still observed this idea has not been followed further.

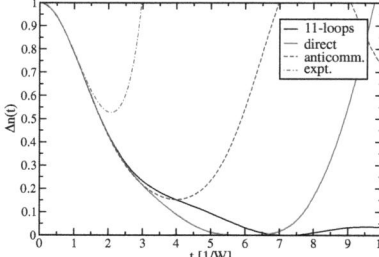

Fig. 4.1.: Jump for a quench to $U = 1.0W$ derived by the different approaches with the result of the 11-loop calculation (solid black line) as reference. Solid red line: Results of a 3-loop calculation. Dashed line: Results of the matrix approach with the scalar product based on the anticommutator [see (4.9)]. Dash-dotted line: Results derived with the scalar product based on the expectation value [see (4.7)].

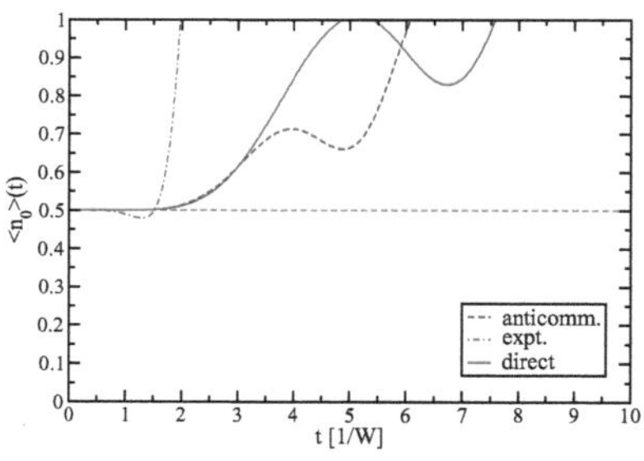

Fig. 4.2.: Corresponding results for the expectation value. The solid line depicts results of a 3-loop calculation. Dashed line: Results of the matrix approach with the scalar product based on the anticommutator as given in Eq. 4.9. Dashed-dotted line: Results derived with the scalar product based on the expectation value due to Eq. 4.7.

4.2. Momentum-Space Approach

An alternative route in determining the jump captures the translational part of the time evolution exactly by performing the calculation in momentum space. The momentum space representation of the creation operator is given by the Fourier transform

$$\hat{c}_k^\dagger = \frac{1}{\sqrt{N}} \sum_j \hat{c}_j^\dagger e^{ikr_j} \tag{4.13}$$

with N denoting the number of sites. To illustrate that the kinetic part is well described by this approach a calculation for the non-interacting system is performed.

With the initial operator Eq. 4.13 the time evolution is calculated by

$$\partial_t \hat{c}_{k,\uparrow}^\dagger(t) = \frac{1}{\sqrt{N}} \sum_j i\left[\hat{H}_0, \hat{c}_{j,\uparrow}^\dagger(t)\right] e^{ikr_j} \tag{4.14a}$$

$$= \frac{1}{\sqrt{N}} \sum_j (Ji)\left(\hat{c}_{j+1,\uparrow}^\dagger(t) + \hat{c}_{j-1,\uparrow}^\dagger(t)\right) e^{ikr_j} \tag{4.14b}$$

$$= -Ji\left(e^{-ik}\hat{c}_{k,\uparrow}^\dagger(t) + e^{ik}\hat{c}_{k,\uparrow}^\dagger(t)\right) \tag{4.14c}$$

$$= -2Ji\cos(k)\hat{c}_{k,\uparrow}^\dagger(t). \tag{4.14d}$$

In the non-interacting case the time dependence of the operator $\hat{c}_{0,k}^\dagger(t)$ can be written as

$$\hat{c}_{0,k}^\dagger(t) = h_{0,k}(t)\hat{c}_k^\dagger \tag{4.15}$$

where the prefactor $h_{0,k}(t)$ representing the time dependence of the one-particle terms is introduced. Taking the initial conditions into account the time evolution is explicitly given by

$$h_{0,k}(t) = e^{-2Ji\cos(k)t}. \tag{4.16}$$

Variants of the Approach

leading to a constant jump $\Delta n = 1$ as expected for the non-interacting model. The expectation value at time t is related to its initial value via

$$\langle 0|\hat{c}^\dagger_{k,\uparrow}(t)\hat{c}_{k,\uparrow}(t)|0\rangle = \langle 0|e^{-2Ji\cos(k)t}\hat{c}^\dagger_{k,\uparrow}e^{2Ji\cos(k)t}\hat{c}_{k,\uparrow}|0\rangle \quad (4.17a)$$

$$= \langle 0|\hat{c}^\dagger_{k,\uparrow}\hat{c}_{k,\uparrow}|0\rangle \quad (4.17b)$$

$$= \hat{n}_k(t=0) \quad (4.17c)$$

demonstrating that the translational part is correctly captured by this approach. A corresponding 2-loop calculation in real space does not capture the translational part correctly (see Sect. 3.4).
In the case of non-vanishing interaction U an additional term is created during the first loop, emerging from

$$\partial_t \hat{c}^\dagger_{k,\uparrow}(t) = \frac{1}{\sqrt{N}}\sum_j i\left[\hat{H}, \hat{c}^\dagger_{j,\uparrow}(t)\right]e^{ikj} \quad (4.18a)$$

$$= -2Ji\cos(k)\hat{c}^\dagger_{k,\uparrow}(t) + Un(1-n)i\frac{1}{\sqrt{N}}\sum_j \hat{c}^\dagger_{j,\uparrow}\hat{c}^\dagger_{j,\downarrow}\hat{c}_{j,\downarrow}(t)e^{ikj} \quad (4.18b)$$

$$= -2Ji\cos(k)\hat{c}^\dagger_{k,\uparrow}(t) + Un(1-n)i\left.\hat{c}^\dagger_\uparrow\hat{c}^\dagger_\downarrow\hat{c}_\downarrow\right|_k(t). \quad (4.18c)$$

The additional monomial $\left.\hat{c}^\dagger_\uparrow\hat{c}^\dagger_\downarrow\hat{c}_\downarrow\right|_k$ created in this commutation denotes a term with a particle and a particle-hole pair. For this monomial an additional prefactor $h_{1,k}(t)$ is used to capture the time dependence of these terms. In the case of non-vanishing interaction the ansatz for the time dependent operator $\hat{c}^\dagger_{k,\uparrow}(t)$ is modified to

$$\hat{c}^\dagger_{k,\uparrow}(t) = h_{0,k}(t)\hat{c}^\dagger_{k,\uparrow} + h_{1,k}(t) : \left.\hat{c}^\dagger_\uparrow\hat{c}^\dagger_\downarrow\hat{c}_\downarrow\right|_k : \quad (4.19)$$

4.2 Momentum-Space Approach

in order to include the three-particle term. The time dependence of this newly created term $\hat{c}^\dagger_\uparrow \hat{c}^\dagger_\downarrow \hat{c}_\downarrow \big|_k$ follows from the commutator

$$\partial_t \hat{c}^\dagger_\uparrow \hat{c}^\dagger_\downarrow \hat{c}_\downarrow \big|_k (t) = \frac{1}{\sqrt{N}} \sum_j i\left[\hat{H}, \hat{c}^\dagger_{j,\uparrow} \hat{c}^\dagger_{j,\downarrow} \hat{c}_{j,\downarrow}(t)\right] e^{ikj} \quad (4.20a)$$

$$= \frac{1}{\sqrt{N}} \sum_j U i \hat{c}^\dagger_{j,\uparrow}(t) e^{ikj} + \frac{1}{\sqrt{N}} \sum_j U(2n-1) i \hat{c}^\dagger_{j,\uparrow} \hat{c}^\dagger_{j,\downarrow} \hat{c}_{j,\downarrow}(t) e^{ikj} \quad (4.20b)$$

$$= U i \hat{c}^\dagger_{k,\uparrow}(t) + U(2n-1) i \hat{c}^\dagger_\uparrow \hat{c}^\dagger_\downarrow \hat{c}_\downarrow \big|_k (t). \quad (4.20c)$$

Comparing both sides of these equations a set of coupled differential equations

$$\partial_t h_{0,k}(t) = -2J i \cos(k) h_{0,k}(t) + U(1-n) n i h_{1,k}(t) \quad (4.21a)$$
$$\partial_t h_{1,k}(t) = U i h_{0,k}(t) + U(2n-1) i h_{1,k}(t) \quad (4.21b)$$

can be deduced. In the half-filled case with $n = 0.5$ and $k = \frac{\pi}{2}$ (and $J = \frac{1}{4}$) these equations are solved by $h_{0,k}(t) = \cos\left(\frac{U}{2}t\right)$ implying an oscillating jump $\Delta n(t) = \cos^2\left(\frac{U}{2}t\right) = \frac{1}{2} + \frac{1}{2}\cos(Ut)$.

4.2.0.2. Expectation Value

To calculate the expectation value of a particle at site 0, the operators $\hat{c}_k$ have to be transformed back into real space by a Fourier transform

$$\hat{c}^\dagger_{j,\uparrow} = \frac{1}{\sqrt{2\pi}} \int_0^\pi dk\, e^{-ikj} \hat{c}^\dagger_{k,\uparrow}. \quad (4.22)$$

Variants of the Approach

Thus the expectation value $\langle 0|\hat{c}_{j,\uparrow}^{\dagger}(t)\hat{c}_{j,\uparrow}(t)|0\rangle$ can be written as

$$\langle 0|\hat{c}_{j,\uparrow}^{\dagger}(t)\hat{c}_{j,\uparrow}(t)|0\rangle = \langle 0|\frac{1}{\pi^2}\int_0^\pi e^{-ikj}\hat{c}_{k,\uparrow}^{\dagger}(t)\mathrm{d}k\int_0^\pi e^{iqj}\hat{c}_{q,\uparrow}(t)\mathrm{d}q|0\rangle \quad (4.23\mathrm{a})$$

$$= \langle 0|\frac{1}{\pi^2}\int_0^\pi\int_0^\pi e^{-i(k-q)j}\hat{c}_{k,\uparrow}^{\dagger}(t)\hat{c}_{q,\uparrow}(t)\mathrm{d}k\mathrm{d}q|0\rangle \quad (4.23\mathrm{b})$$

which simplifies for $j = 0$ to

$$\langle 0|\hat{c}_{0,\uparrow}^{\dagger}(t)\hat{c}_{0,\uparrow}(t)|0\rangle = \frac{1}{\pi^2}\langle 0|\int_0^\pi\int_0^\pi \hat{c}_{k,\uparrow}^{\dagger}(t)\hat{c}_{q,\uparrow}(t)\mathrm{d}k\mathrm{d}q|0\rangle \quad (4.24\mathrm{a})$$

$$= \frac{1}{\pi}\int_0^\pi \langle 0|\hat{c}_{k,\uparrow}^{\dagger}(t)\hat{c}_{k,\uparrow}(t)|0\rangle\mathrm{d}k. \quad (4.24\mathrm{b})$$

For the 2-loop calculation this leads to

$$\langle 0|\hat{c}_{k,\uparrow}^{\dagger}(t)\hat{c}_{k,\uparrow}(t)|0\rangle = n|h_0|^2 + (1-n)\cdot n^2|h_1|_k^2 \quad (4.25)$$

which yields the exact value $n_0 = \frac{1}{2}$. This correspondence to the exact result is coincidental and due to the fact, that only two loops are performed.

4.2.0.3. Results for 3 Loops

For a higher number of loops the differential equations have to be solved numerically. Results for a 3-loop calculation in momentum space with $U = 1.0W$ are depicted in Fig. 4.3. For comparison the corresponding real space calculation is shown additionally. Taking the curve for a real space calculation with 11 loops as reference, both curves deviate at the same time t. Furthermore both curves shoot up to unphysical values for larger times.

4.2 Momentum-Space Approach

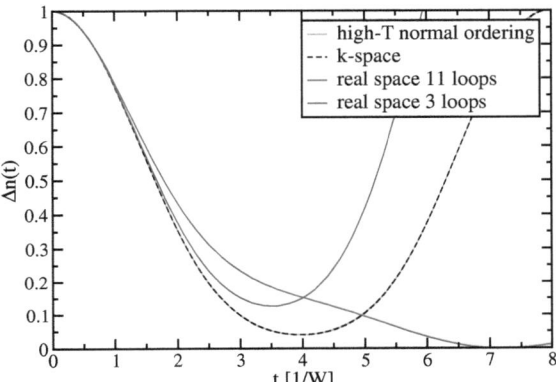

Fig. 4.3.: Jump $\Delta n(t)$ for a quench to $U = 1.0W$ derived by calculations in real and momentum space with three loops, with the result of the 11-loop calculation given as reference.

To check whether the non-unitarity is reduced by the use of the momentum space approach the local expectation value is depicted for the two 3-loop calculations in Fig. 4.4. The curves deviate from their initial value at about the same time t. Thus the momentum space method also shows effects of non-unitarity. The convergence is comparable, but the evaluation of the expectation value is much more demanding in the momentum space approach since many terms have to be considered in the Fourier transform of the operator product in Eq. 4.23b.

To check in how far the results are influenced by the choice of the normal ordering, another type of normal ordering is applied. This normal ordering corresponds to normal ordering with respect to a state with high temperature. In this approach only local pairings yield non-vanishing effects. This approach can also be understood as a calculation with a different choice for the scalar product as explained in Sect. 4.1. The corresponding scalar product for two op-

Variants of the Approach

erators $\hat{A}$ and $\hat{B}$ is given by

$$\left(\hat{A}|\hat{B}\right) = \frac{\text{Tr}(\hat{A}^\dagger \hat{B})}{\text{Tr}\mathbb{1}}. \tag{4.26}$$

The results for a 3-loop calculation by the use of this normal ordering are shown in Fig. 4.5.

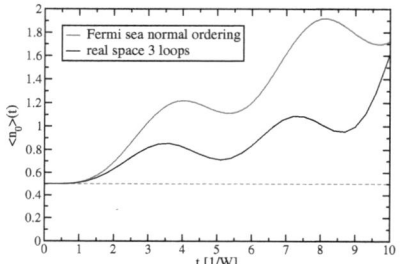

Fig. 4.4.: Local expectation value $\langle \hat{n}_0 \rangle(t)$ for $U = 1.0W$ derived in calculations in real and momentum space with three loops.

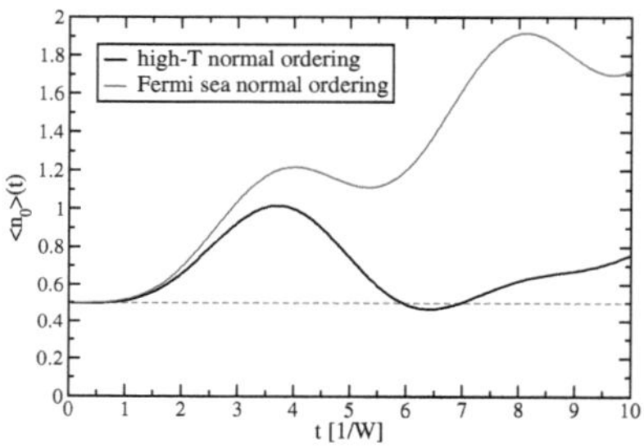

Fig. 4.5.: Local expectation value for momentum space calculations with three loops and $U = 1.0W$. The curves represent results derived by a calculation applying normal ordering with respect to the Fermi sea and normal ordering with respect to high temperature states.

For the jump the approaches yield similar results. The curves lie above each other. A corresponding curve for the jump derived by the use of this type of normal ordering is given in Fig. 4.3 as solid, red line. This curve coincides with the curve obtained applying normal ordering with respect to the Fermi sea. The results for the expectation value yield differences. In the case of normal ordering due to high temperatures the curve stays rather low, whereas the curve for the normal ordering due to the Fermi sea shoots up. However, the range of convergence of the two curves is comparable.

4.2 Momentum-Space Approach

As even these two very distinct types of normal ordering yield similar results it is not expected that the results can essentially be improved by choosing a different type of normal ordering.

Due to the additional Fourier transform needed to determine expectation values a calculation performed in this way is much more demanding than a real space calculation. As the momentum space approach does not resolve the problem of non-unitary results and yields about the same range of convergence in a much more demanding calculation this approach is not followed further.

4.3. Self-Similar Calculations in U^2

4.3.1. Non-Interacting Case

To gain first insights into the behavior of the system after a quench the time evolution of the jump is calculated in a first approach up to second order in the interaction U. This calculation is proportional to U^2 and obtained in momentum space. In this way it captures the non-interacting part exactly.

Since also higher orders in U are created during the calculation, this approach differs from the second order results obtained by Moeckel and Kehrein [46], where a strict U^2 calculation is used. For the two-dimensional model a strict U^2-calculation following Moeckel and Kehrein is introduced in Sect. 6.1.5. In the strict U^2-calculation no higher orders in U appear.

In contrast to this the calculation presented in the following is a full calculation in the subspace of operators describing one and three particles. These are the monomials appearing in second order in U, but during the calculation also higher orders in U are created.

The interaction term can be written in the form

$$U\sum_i :\hat{n}_{i,\uparrow}\hat{n}_{i,\downarrow}: = U\sum_i \left(\hat{n}_{i,\uparrow}-\frac{1}{2}\right)\left(\hat{n}_{i,\downarrow}-\frac{1}{2}\right) \quad (4.27a)$$

$$= \frac{U}{N}\sum_{k_1,k_2,q}\left(\hat{c}^\dagger_{k_1+q,\uparrow}\hat{c}_{k_1,\uparrow}-\delta_{0,\vec{q}}\Theta(k_F-|k_1|)\right)$$

$$*\left(\hat{c}^\dagger_{k_2-q,\downarrow}\hat{c}_{k_2,\downarrow}-\delta_{0,\vec{q}}\Theta(k_F-|k_2|)\right) \quad (4.27b)$$

with the dispersion $\epsilon(k) = -\frac{W}{2}\cos(k) = -2J\cos(k)$. The time evolution is calculated by the Heisenberg equation, which can be split into

$$\partial_t \hat{c}^\dagger_{k,\uparrow}(t) = i\left[\hat{H}, \hat{c}^\dagger_{k,\uparrow}(t)\right] \quad (4.28a)$$

$$= i\underbrace{\left[\hat{H}_0, \hat{c}^\dagger_{k,\uparrow}(t)\right]}_{=:\mathcal{L}_0 \hat{c}^\dagger_{k,\uparrow}(t)} + i\underbrace{\left[\hat{H}_{int}, \hat{c}^\dagger_{k,\uparrow}(t)\right]}_{=:\mathcal{L}_{int}\hat{c}^\dagger_{k,\uparrow}(t)} \quad (4.28b)$$

4.3 Self-Similar Calculations in U^2

with the non-interacting part

$$\mathcal{L}_0 \hat{c}^\dagger_{k,\uparrow}(t) = \epsilon_k \hat{c}^\dagger_{k,\uparrow}(t). \tag{4.29}$$

In the non-interacting case $U = 0$ the time evolution is given through

$$\hat{c}^\dagger_{k,\uparrow}(t) = e^{i\epsilon_k t} \hat{c}^\dagger_{k,\uparrow}(0). \tag{4.30}$$

The real space representation is calculated through a Fourier transform

$$\hat{c}^\dagger_{j,\uparrow}(t) = \frac{1}{N} \sum_k e^{i\epsilon_k t} \sum_j e^{ir_j k} \hat{c}^\dagger_{j,\uparrow}(0) \tag{4.31a}$$

$$= \sum_j h_j(t) \hat{c}^\dagger_{j,\uparrow}(0) \tag{4.31b}$$

with the time dependent prefactor $h_j(t)$. For the continuous representation of the prefactors the following identity holds

$$h_j(t) = \frac{1}{2\pi} \int_{-\pi}^{\pi} e^{i(\epsilon_k t + r_j k)} dk \tag{4.32a}$$

$$= \frac{1}{\pi} \int_0^\pi e^{i2Jt\cos(k)} \cos(jk) dk \tag{4.32b}$$

$$\tag{4.32c}$$

where the sine term is omitted due to symmetry. Thus the prefactor finally reads

$$h_j(t) = (-i)^j J_j(2Jt) \tag{4.33a}$$

$$= (-i)^j J_j\left(\frac{W}{2} t\right). \tag{4.33b}$$

with the Bessel function $J_j(x)$.

Variants of the Approach

4.3.2. Interacting Case

4.3.2.1. Setting up the differential equations

Applying the Liouville operator $\mathcal{L}_{\text{int}}$ to the creation operator leads to

$$\mathcal{L}_{\text{int}} \hat{c}^\dagger_{k,\uparrow}(t) = \frac{U}{N} \sum_{k_1,k_2,q} \underbrace{\left[:\hat{c}^\dagger_{k_1+q,\uparrow} \hat{c}_{k_1,\uparrow} :, \hat{c}^\dagger_{k,\uparrow} \right]}_{\delta_{k_1,k} \hat{c}^\dagger_{q+k_1,\uparrow}} :\hat{c}^\dagger_{k_2-q,\downarrow} \hat{c}_{k_2,\downarrow}: \quad (4.34a)$$

$$= -\frac{U}{N} \sum_{k_2,q} :\hat{c}^\dagger_{k+q,\uparrow} \hat{c}^\dagger_{k_2-q,\downarrow} \hat{c}_{k_2,\downarrow}: \quad (4.34b)$$

where in the normal ordered term the operators with a spin pointing downwards are split off from the ones with a spin pointing up. This is justified by the normal ordering, as operators with opposite spin do not yield contractions. The term $:\hat{c}^\dagger_{k+q,\uparrow} \hat{c}^\dagger_{k_2-q,\downarrow} \hat{c}_{k_2,\downarrow}:$ is newly created by the commutation and has to be considered in the following calculations. For the results to be correct up to second order in U the Liouvillian has to be applied once more to the terms created during the first commutation. The calculation of the commutator can be found in App. A.

To describe the translational part of the time evolution of the monomial $:\hat{c}^\dagger_{k+q,\uparrow} \hat{c}^\dagger_{k_2-q,\downarrow} \hat{c}_{k_2,\downarrow}:$ given analogously to Eq. 4.30 the energy difference of the operators contained in the monomial is needed. For the three-particle term the energy difference $d_{k_2,k,q} = \epsilon_{k+q} + \epsilon_{k_2-q} - \epsilon_{k_2}$ is introduced.

With $h_{k_2,k,q}$ denoting the relevant combinations of the particle numbers for the momenta involved

$$h_{k_2,k,q} = -n_{k_2-q} n_{k_2} + n_{k+q} n_{k_2} + n_{k+q} n_{k_2-q} \quad (4.35)$$

an additional term

$$\left[\hat{H}_{\text{int}}, :\hat{c}^\dagger_{k+q,\uparrow} \hat{c}^\dagger_{k_2-q,\downarrow} \hat{c}_{k_2,\downarrow}: \right] = i\frac{U}{N} h_{k_2,k,q} :\hat{c}^\dagger_{k,\uparrow}: \quad (4.36)$$

66

4.3 Self-Similar Calculations in U^2

is obtained by the commutator.

Interpreting the prefactor of the one-particle terms $\hat{c}^\dagger_{k_F,\uparrow}$ as first component v_0 and the prefactors of the three-particle terms : $\hat{c}^\dagger_{k_F+q,\uparrow}\hat{c}^\dagger_{k_2-q,\downarrow}\hat{c}_{k_2,\downarrow}$: as the following entries $v_{k_2,k_F,q}$ of a vector $\vec{v}(t)$ the differential equations can be summarized in a matrix equation

$$\partial_t \vec{v}(t) = i \begin{pmatrix} \epsilon_F & U/N h_{k_2,k_F,q} & \cdots \\ U/N & \ddots & & 0 \\ \vdots & & d_{k_2,k_F,q} & \\ & 0 & & \ddots \end{pmatrix} \vec{v}(t) \quad \text{with} \quad \vec{v}(t=0) = \begin{pmatrix} 1 \\ 0 \\ \vdots \\ 0 \end{pmatrix} \tag{4.37}$$

for the initial vector. The time derivative of the prefactors for the three-particle terms can be deduced

$$\partial_t v_{k_2,k_F,q} = i d_{k_2,k_F,q} v_{k_2,k_F,q} + i\frac{U}{N} v_0 \tag{4.38}$$

which can be multiplied by $e^{id_{k_2,k_F,q}t}$ to yield

$$\partial_t \left(v_{k_2,k_F,q} e^{id_{k_2,k_F,q}t} \right) = i\frac{U}{N} v_0 e^{id_{k_2,k_F,q}t} \tag{4.39a}$$

$$\Rightarrow v_{k_2,k_F,q} = i\frac{U}{N} \int_0^t v_0(t') e^{-id_{k_2,k_F,q}(t-t')} dt'. \tag{4.39b}$$

4.3.2.2. Calculating the Green function

The closed form of the differential equations leads to a self-energy equation as is shown next. With the Fermi energy $\epsilon_F = 0$ the differential equation for the prefactor v_0 is given through

$$\partial_t v_0(t) = i\frac{U}{N}\sum_{k_2,q} h_{k_2,k_F,q} v_{k_2,k_F,q}$$
$$-\frac{U^2}{N^2}\sum_{k_2,q} h_{k_2,k_F,q} \int_0^t v_0(t') e^{id_{k_2,k_F,q}(t-t')} dt'. \qquad (4.40)$$

The prefactor $v_0(t)$ has to be proportional to a theta function $v_0(t) \propto \Theta(t)$ as the prefactor has to obey $v(t < 0) = 0$. Using the identity

$$\mathcal{G}(t) = -i\Theta(t)\frac{1}{N^2}\sum_{k_2,q} h_{k_2,k_F,q} e^{id_{k_2,k_F,q}t} \qquad (4.41)$$

the time derivative of the prefactor

$$\partial_t v_0(t) = -iU^2 \int_{-\infty}^{\infty} v_0(t') \mathcal{G}(t-t') dt' + \delta(t) \qquad (4.42)$$

follows where the δ-function ensures that the initial conditions are fulfilled.
Through a Fourier transformation

$$v_0(t) = \frac{1}{2\pi} \int_{-\infty}^{\infty} v_0(\omega) e^{-i\omega t} d\omega \qquad (4.43a)$$

$$\mathcal{G}(t) = \frac{1}{2\pi} \int_{-\infty}^{\infty} \mathcal{G}(\omega) e^{-i\omega t} d\omega \qquad (4.43b)$$

4.3 Self-Similar Calculations in U^2

are introduced, which can be inserted to solve Eq. 4.42

$$\int_{-\infty}^{\infty}(-i\omega)v_0(\omega)e^{-i\omega t}d\omega$$
$$= -U^2 i \int_{-\infty}^{\infty} d\omega v_0(\omega)\mathcal{G}(\omega)e^{-i\omega t} + \int_{-\infty}^{\infty} e^{-i\omega t}d\omega \quad (4.44)$$

from which

$$-i\omega v_0(\omega) = -U^2 i v_0(\omega)\mathcal{G}(\omega) + 1 \quad (4.45a)$$
$$\Leftrightarrow v_0(\omega) = \frac{i}{(\omega - U^2\mathcal{G}(\omega))} \quad (4.45b)$$

can be concluded. The time dependence of $v_0(t)$ determining the jump $\Delta n(t) = |v_0(t)|^2$ is given by a Fourier transformation of Eq. 4.45b. The resulting self-energy equation reads

$$v_0(t) = \frac{1}{2\pi}\int_{-\infty}^{\infty}\frac{i}{\omega - U^2\mathcal{G}(\omega)}e^{-i\omega t}d\omega. \quad (4.46)$$

Since a term $\propto U^2$ appears in the denominator this term is $O(U^2)$. Obviously also higher orders in U enter the calculation. The Hilbert representation can be written as

$$\frac{1}{\omega - U^2\mathcal{G}(\omega)} = \int_{-\infty}^{\infty}\frac{\rho(x)dx}{\omega - x} \quad (4.47)$$

with the positive spectral density $\rho(x) \geq 0$.

Variants of the Approach

By the residue theorem the time dependence of v_0 satisfies

$$v_0(t) = \frac{i}{2\pi} \int_{-\infty}^{\infty} dx \rho(x) \oint_\Gamma \frac{e^{-i\omega t}}{\omega - x + i0^+ t} d\omega \qquad (4.48a)$$

$$= \int_{-\infty}^{\infty} dx \rho(x) e^{-ixt}. \qquad (4.48b)$$

Thus the spectral density $\rho(x)$ belonging to $\frac{i}{\omega - U^2 \mathcal{G}(\omega)}$ is needed to calculate the time dependence of $v_0(t)$.

In a next step the function $\mathcal{G}(\omega)$ is calculated under the condition $\mathrm{Im}\,\omega = 0^+$

$$\mathcal{G}(\omega) = \int_0^\infty e^{i\omega t} dt \frac{-i}{(2\pi)^2} \int_{-\pi}^{\pi} \int dk_2 dq h_{k_2,k_F,q} e^{id_{k_2,k_F,q} t} \qquad (4.49a)$$

$$= \frac{-i}{(2\pi)^2} \int_{-\pi}^{\pi} \int dk_2 dq h_{k_2,k_F,q} \underbrace{\int_0^\infty e^{i(\omega + d_{k_2,k_F,q})t} dt}_{\left. \frac{e^{i(\omega + d_{k_2,k_F,q})t}}{i(\omega + d_{k_2,k_F,q})} \right|_0^\infty}. \qquad (4.49b)$$

The asymptotic behavior of the Euler function is determined through the imaginary part of ω and $d_{k_2,k_F,q}$. As $\mathrm{Im}\,\omega \to 0^+$ and $\mathrm{Im}\,d_{k_2,k_F,q} = 0$, the asymptotic behavior implies

$$\mathcal{G}(\omega) = \frac{-i}{(2\pi)^2} \int_{-\pi}^{\pi} \int dk_2 dq h_{k_2,k_F,q} \frac{i}{\omega + d_{k_2,k_F,q}} \qquad (4.50)$$

4.3 Self-Similar Calculations in U^2

which can be rewritten

$$\mathcal{G}(\omega) = \frac{-i}{(2\pi)^2} \int\int_{-\pi}^{\pi} dk_2 dq\, h_{k_2,k_F,q} \frac{i}{\omega + d_{k_2,k_F,q}} \tag{4.51a}$$

$$= \frac{1}{(2\pi)^2} \int_{-\infty}^{\infty}\int_{-\pi}^{\pi}\int_{-\pi}^{\pi} dx\, dk_2 dq\, \delta(x + d_{k_2,k_F,q}) h_{k_2,k_F,q} \frac{\rho(x)}{\omega - x} \tag{4.51b}$$

$$= \frac{1}{(2\pi)^2} \int_{-\pi}^{\pi}\int_{-\pi}^{\pi} dx\, dk_2 dq\, h_{k_2,k_F,q} \frac{\rho(-d_{k_2,k_F,q})}{\omega - x} \tag{4.51c}$$

$$= \int_{-\infty}^{\infty} \frac{\rho_G(x)}{\omega - x} dx \tag{4.51d}$$

with the spectral function

$$\rho_G(x) = \frac{1}{(2\pi)^2} \int\int_{-\pi}^{\pi} dk_2 dq\, \delta(x + d_{k_2,k_F,q}) h_{k_2,k_F,q} \geq 0. \tag{4.52}$$

The spectral density can be split into two parts $\rho_G(x) = \rho_+(x) + \rho_-(x)$ for positive and negative argument. The resulting Green function is given by

$$\mathcal{G}(\omega) = \int_0^{\infty} \frac{\rho_+(x)}{\omega - x} dx + \int_{-\infty}^0 \frac{\rho_-(x)}{\omega - x} dx \tag{4.53}$$

where the first part fulfills $\rho_+(x) > 0$ for $x > 0$ and $\rho_+(x) = 0$ for $x < 0$. The second part obeys $\rho_-(x) > 0$ for $x < 0$ and $\rho_-(x) = 0$ for $x > 0$. The explicit calculation of the spectral density is given in App. A.2. In this section only the results are recalled.

Variants of the Approach

As explained in App. A.2 the spectral density can be calculated numerically by finding the zeros of the argument of a δ-function and evaluating the integral determining the spectral density numerically.

With $\rho_-(x)$ and $\rho_+(x)$ the Green function $\mathcal{G}(\omega)$ can be calculated by Eq. 4.53. Results for the Green function are shown in Fig. 4.6.

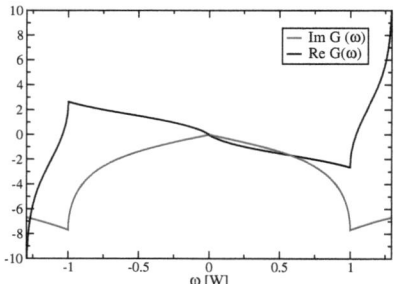

Fig. 4.6.: Real and imaginary part of the Green function $\mathcal{G}(\omega)$ in dependence on ω.

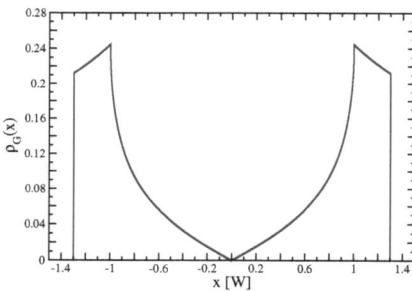

Fig. 4.7.: Spectral density $\rho_G(x)$ as defined by the Hilbert representation of the Green function $\mathcal{G}(\omega)$.

The spectral density $\rho_G(x)$ is shown in Fig. 4.7. For small $|x|$ the spectral density shows a linear increase. The linear behavior is due to the one-dimensionality of the model. In a real Fermi liquid $\rho_G(x)$ would increase quadratically for small $|x|$. The reduced dimensionality hampers the quadratic increase, leading to a linear behavior.

Besides, the spectral density shows band edges.

As two momenta are involved in the calculation these band edges are analogous to van-Hoove singularities in two-dimensional models, i.e., they are $O(x^0)$ and thus the depicted jumps occur.

For $|x| = 1.0$ the spectral density for the one-dimensional model exhibits square root singularities.

The model does not show infrared singularities as an infrared cutoff $\propto \frac{1}{t}$ exists.

4.3 Self-Similar Calculations in U^2

With $\mathcal{G}(\omega)$ the Hilbert representation of $v_0(t)$ can be calculated by Eq. 4.47.

To determine the time dependence $v_0(t)$ the spectral density $\rho(x)$ has to be calculated. It is shown for various values of the interaction U in Fig. 4.8.

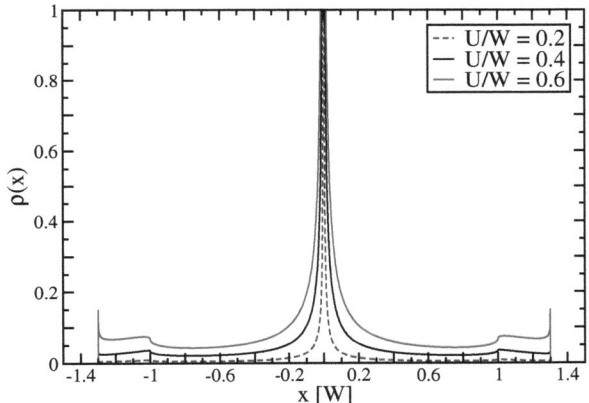

Fig. 4.8.: Density $\rho(x)$ in dependence on x for various interaction strengths U. For small $|x|$ the density exhibits a sharp peak with an increase $\propto \frac{1}{x\ln(x)^2}$.

The linear increase observed for small $|x|$ in $\rho_G(x)$ leads to a peak in the spectral density $\rho(x)$. For small $|x|$ the density $\rho(x)$ is proportional to $\frac{1}{x\ln(x)^2}$ in the one-dimensional model. In a real Fermi liquid the quadratic increase in $\rho_G(x)$ would result in a δ-peak in $\rho(x)$. For larger $|x|$ the density shows cutoffs with resonances. Besides, the spectral function exhibits delta-peaks for larger $|x|$. These are signatures of trions given as bounded states of two particles and a hole. These states are more important for large interactions U. Thus the peaks appear more pronounced for

Variants of the Approach

increasing U.

The time dependence $v(t)$ is then given by a Fourier transform of $\rho(x)$. With $v_0(t)$ the behavior of the jump $n_t(k) = n_0(k)|v_0(t)|^2$ can be determined. The jump $\Delta n(t)$ for various values of the interaction U is shown in Fig. 4.9 and Fig. 4.10. In these curves the interaction U is increased from top to bottom.

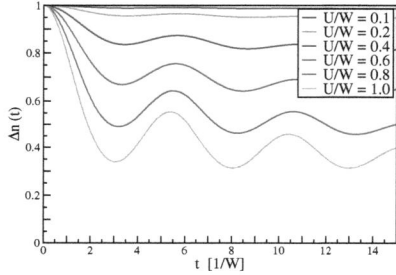

Fig. 4.9.: Jump $\Delta n(t)$ in dependence on the time t derived in the strict U^2 calculation for various values of the quench depths U (increasing U from top to bottom).

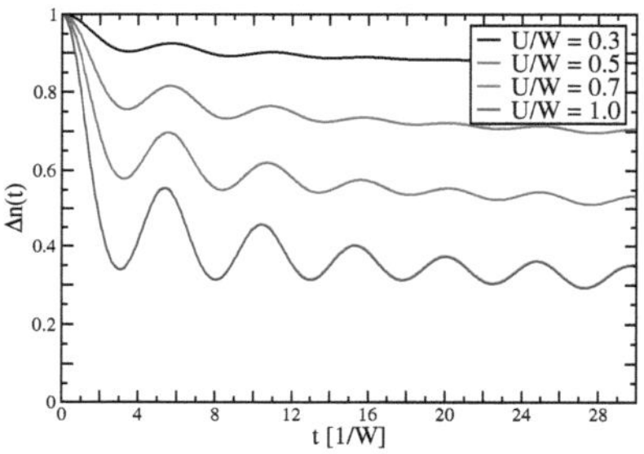

Fig. 4.10.: Jump $\Delta n(t)$ for various values of U shown over longer times. On increasing U (from top to bottom) a stronger reduction of the jump is observed.

With increasing interaction strength U the decrease in the jump becomes stronger. The oscillations in the time dependence result from the band edges in $\rho_G(x)$. As these are more pronounced for larger U the amplitudes of the oscillations are increased on increasing U. For large values of the interaction U the three-particle states become more important. This supports the view that the oscillations are to be attributed to the band edges.

For these rather large interaction values U higher orders in U have to be considered to capture all relevant processes.

For the second order calculation only two momenta are required, which makes it feasible.

4.3 Self-Similar Calculations in U^2

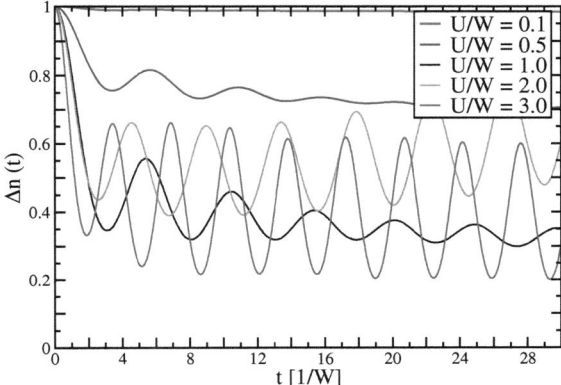

Fig. 4.11.: Jump for various interaction strengths. It can be seen how the amplitude of the oscillations is increased on increasing U.

Of course a calculation which is exact up to third order U^3 would be possible, but such a calculation would require to include more momenta. Consequently such a calculation would be very demanding. Thus it is advantageous to use a more direct approach in real space as used in the following.

4.3.3. Comparison to the 11-Loop Calculation

In Fig. 4.12 the results of the U^2-calculation (dashed lines) are compared to the results of an 11-loop calculation (solid lines) for various interaction strengths U.

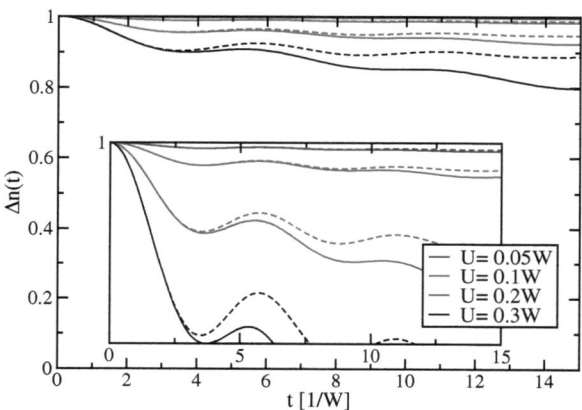

Fig. 4.12.: Comparison of the jump as given by the U^2-calculation according to Sect. 4.3 (dashed lines) with the results of the 11-loop calculation (solid lines) for various values of the interaction U (with increasing U from top to bottom).

Even for small values of the interaction the results deviate already for small times. For instance for $U = 0.2W$ the U^2 result deviates from the 11-loop result at $t \approx 4/W$. Thus the results of the U^2-calculation are only reliable over a very small range in time. The processes which govern the behavior of the jump beyond the first drop are not described by the U^2 approach. Consequently higher orders have to be included as it is the case in the equation of motion approach.

5. Results for the One-Dimensional Model

The results of the calculation in second order in the interaction U presented in Sect. 4.3 are restricted to small interactions U. As explained these results deviate rather early from the exact results. Besides, a generalization to higher orders in U is not straightforward. Consequently the iterated equation of motion approach is used in the following to describe the dynamics of the system. In this way also higher orders in the interaction U are captured.

In this chapter results for the one-dimensional Hubbard model derived in the iterated equation of motion approach are presented.

A self-similar truncation as explained in Sect. 3.4 is applied to improve the convergence. In the following the term **loop** denotes such a self-similar calculation within the iterated equation of motion approach. In this sense an m-loop calculation has to be understood as a calculation with m commutations where the monomials appearing in the last loop for the first time are neglected.

5.1. Half-Filled Model

In the following results for the half-filled model with $n = 0.5$ denoting the filling factor for one spin species are discussed.

5.1.1. Jump for Various Interaction Strengths U

Results for the jump $\Delta n(t)$ for various values of the interaction strength U are depicted in Fig. 5.1. The dependence of the decrease of the jump on the interaction U can be seen. Besides, the decreased ranges of convergence for larger quenches are visible. Corresponding results for stronger quenches can be

Results for the One-Dimensional Model

found in Sect. 5.4.1. The results presented here are derived in a calculation with 11 loops. In Fig. 5.1 the solid part of the curves represents results for the time scales where the results are converged according to the criteria discussed in Sect. 3.2.2. In contrast to this the dashed part of the curves represent results for times beyond the range of convergence. Consequently this part of the curves has to be treated cautiously.

It can be seen that especially for large U the dashed lines increase rather quickly towards unphysical values larger than 1. This behavior is to be attributed to the breakdown of the unitarity in the equation of motion approach for too large times t, cf. Sect. 4.1.

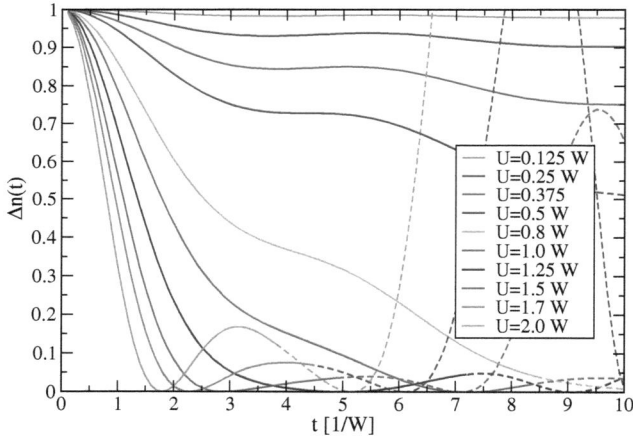

Fig. 5.1.: Jump $\Delta n(t)$ for the half-filled Hubbard model for various values of the interaction strength U as function of time t. The results are obtained by an 11-loop calculation. The dashed part of the curves presents results which lie beyond the range of convergence and may not be considered reliable. The curves are shown for values of $U = 0.125$ to $U = 2.0W$ (from top to bottom for small t).

5.1 Half-Filled Model

In the range where the results are converged the curves represent the exact dynamics of the jump. The figure illustrates how the jump is decreased on increasing U. The curves depict oscillations stemming from the band edges of the dispersion, which appear already in a self-similar U^2-calculation as presented in Sect. 4.3.
The change from a gradually decreasing jump for small U to a jump showing pronounced oscillations for large U can be observed in Fig. 5.2. In this figure the jump is shown in dependence on the time t and the interaction strength U. The curves are shown in the time range where they are converged.
On passing from a gradually decreasing jump to a jump with pronounced oscillations the system crosses a dynamical transition discussed in Sect. 5.4.2.

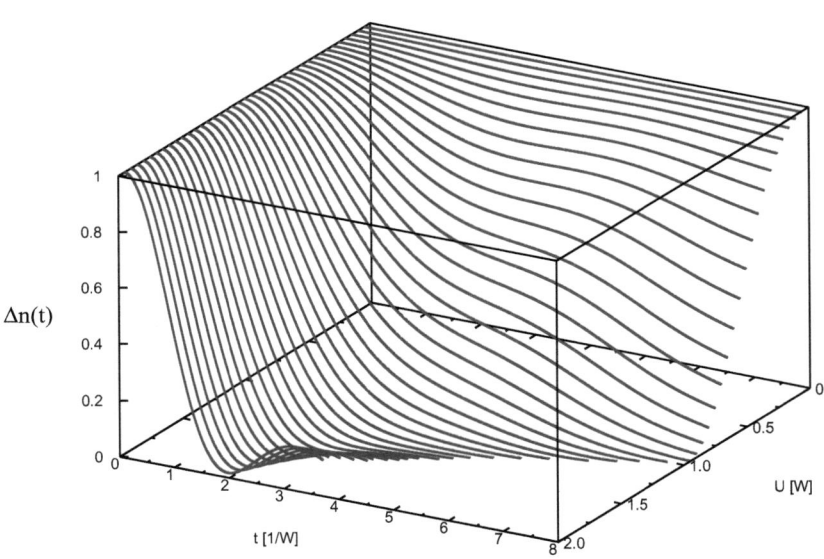

Fig. 5.2.: Time evolution of the jump $\Delta n(t)$ in dependence on the interaction strength U for values from $U=0$ to $U=2.0W$ as derived in a 9-loop calculation. The strong U-dependence of the jump is illustrated.

79

Results for the One-Dimensional Model

5.1.2. Momentum Distribution

The whole momentum distribution (MD) is derived by a numerical Fourier transform of the expectation value $\langle 0|\hat{c}^\dagger_{r,\uparrow}(t)\hat{c}_{0,\uparrow}(t)|0\rangle$ defined by

$$n_k(t) = \frac{1}{N}\sum_r \langle 0|\hat{c}^\dagger_{r,\uparrow}(t)\hat{c}_{0,\uparrow}(t)|0\rangle e^{ikr}. \tag{5.1}$$

In principle these expectation values involve the products of all terms included in the derivation of the differential equation. But since expectation values of two monomials with different numbers of operators or different numbers of operators with a spin pointing upwards vanish, the number of neccessary products is reduced. Nevertheless the exponentially growing number of monomials restricts the derivation of the complete momentum distribution to calculations with seven loops.

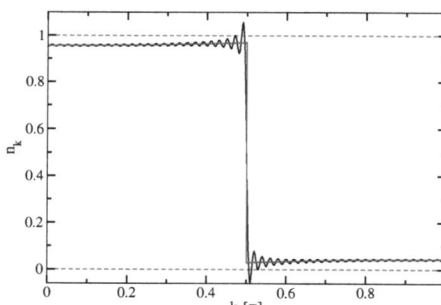

Fig. 5.3.: Black line: Momentum distribution for a quench to $U = 1.0W$ and a time $t = 0.5/W$ derived in a 7-loop calculation. Red line: Results for the momentum distribution where the oscillations due to the Gibbs phenomenon are substracted.

The results of a calculation with about a hundred k-values for a quench to $U = 1.0W$ and $t = 0.5/W$ after the quench are shown in Fig. 5.3 in dependence on the momentum k. The momentum distribution is overlayed by oscillations which can be seen as artefacts of the Fourier transform. These oscillations can be avoided as explained below.

At half-filling the momentum distribution exhibits a jump at

5.1 Half-Filled Model

$k = k_F = 0.5\pi$. When applying the Fourier transform this discontinuity leads to the Gibbs phenomenon, as observed by the oscillations overlying the jump. The Gibbs phenomenon can be suppressed by considering the discontinuity separately in the Fourier transform. The discontinuity is caused by the one-particle terms but it is not the only effect of the one-particle terms. The Fourier coefficients for the jump can easily be calculated analytically. Due to the symmetry the Fourier transform is given in terms of the cosine-coefficients, which read

$$b_j = \frac{2}{\pi} \frac{\sin(kj)}{j} \cdot \Delta n(t) \quad \forall j \neq 0$$
$$b_0 = k\Delta n(t). \tag{5.2}$$

Since the Fourier coefficients of the terms causing the jump are known exactly, the oscillations can be avoided by performing an analytical Fourier transform for these terms and a numerical Fourier transform for the remaining terms. Thus the part of the one-particle terms causing the jump is subtracted before the Fourier transform. After a numerical Fourier transform of the remaining terms the jump is added so that the oscillations caused by the Gibbs phenomen are avoided. The red curve in Fig. 5.3 depicts the momentum distribution adjusted by the oscillations due to the discontinuity at $k = \pi/2$. These results are derived as explained treating the jump separately. As the momentum distribution is shown for a time of $t = 0.5/W$ and a relatively large interaction strength $U = 1.0W$ the jump $\Delta n(t)$ is already reduced. The momentum distribution starts for $k = 0$ below one. Besides it shows a curvature leading to an increase of the momentum distribution towards k_F for $k < k_F$. To illustrate how the curvature develops in time, the momentum distribution for a quench to $U = 1.0W$ and various times is plotted in Fig. 5.4. Clearly the reduction of the jump becomes stronger with increasing time. For a fixed momentum k the momentum distribution evolves non-monotonically, showing oscillations which has also been observed in other calculations [26,46,72].

Results for the One-Dimensional Model

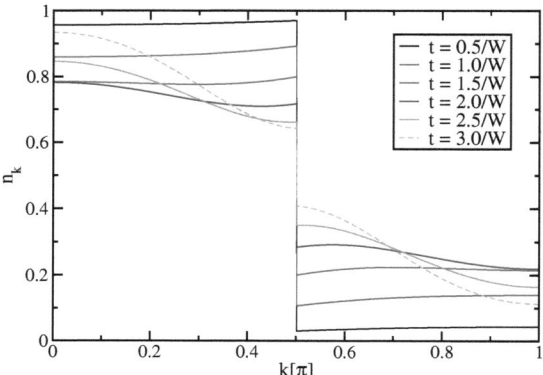

Fig. 5.4.: Momentum distribution for a quench to $U = 1.0W$ derived in a 7-loop calculation. The curves represent the result for various times t (from top to bottom for $k \approx k_F$). On increasing time the jump $\Delta n(t)$ is reduced further.

But the curvature behaves differently from the one observed in the Tomonaga-Luttinger model. The momentum distribution in the Tomonaga-Luttinger model depicts a pure decrease on increasing k for small times [72]. In the Hubbard model the momentum distribution is increased for $k < k_F$ on increasing k for small times after the quench. On the other side of the jump ($k > k_F$) the momentum distribution is increased towards larger k. For larger times (for instance $t = 2.0/W$) the momentum distribution is first decreased and then slightly increased towards k_F. This undulatory behavior is pronounced for larger times.

The evolution of the momentum distribution towards the oscillatory behavior can be seen in Fig. 5.5. The results are shown for a quench to $U = 1.0W$ so that the jump is quickly reduced on increasing time. Focussing on the results for a fixed k, for instance at $k = \pi$, oscillations in time can be observed. Directly at the Fermi surface the curves also show oscillations in time t.

5.1 Half-Filled Model

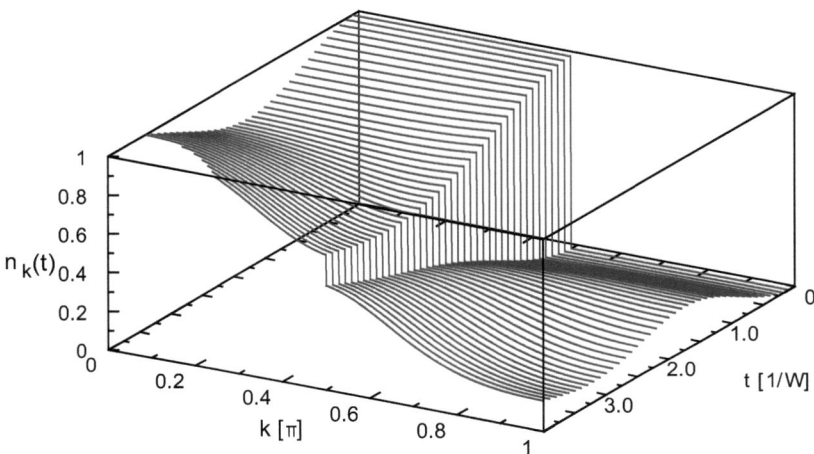

Fig. 5.5.: Momentum distribution for a quench to $U = 1.0W$ in dependence on the time t. The fast decrease of the jump can be observed. On increasing time the momentum distribution exhibits oscillatory behavior.

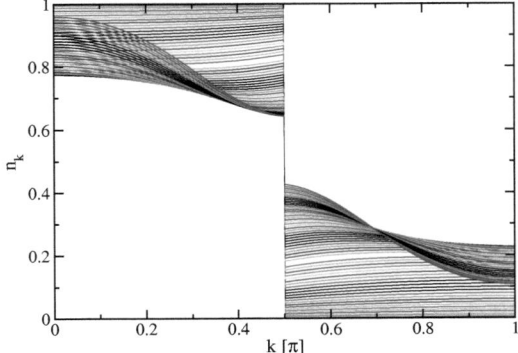

A comparison of the curves for various times is given in Fig. 5.6. It can be observed how the oscillatory behavior of the momentum distribution is built up on increasing t. The momentum distribution changes from relatively flat curves for small times to curves showing oscillations that become stronger on increasing time.

Fig. 5.6.: Momentum distribution for a quench to $U = 1.0W$ for various times from 0.0 to $3.15/W$ in steps of $0.05/W$ (from top to bottom for $k \approx k_F$ with $k < k_F$). The built-up of the oscillatory behavior for larger times is clearly visible.

Results for the One-Dimensional Model

On increasing t the curvature of the momentum distribution increases on both sides of the jump.
For too large times $t \gtrsim 3.0/W$ the momentum distribution is shifted for all k upwards towards larger values. This breaks the particle-hole symmetry of the model, as the curve for $t = 3.0/W$ starts at $k = 0$ with 0.93 and ends at $k = \pi$ with 0.11.
This signals the breakdown of the calculation for such long times.

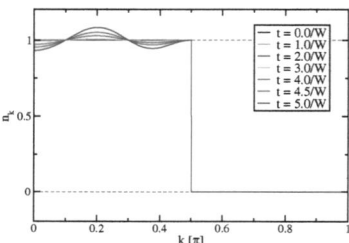

Fig. 5.7.: Momentum distribution for a vanishing interaction U obtained in a 5-loop calculation for various observation times t.

A study of the range of convergence reveals that the times where this shift appears are beyond the range of convergence of the 7-loop calculation. The momentum distribution for the interaction-free case as derived in a 5-loop calculation is depicted in Fig. 5.7 for various times t. For small times the momentum distribution still takes the box shape as it should be.

For $t \approx 4.0/W$ the curves start to deviate. On increasing time larger deviations from this shape are observable for $k < k_F$.
For large times the momentum distribution exhibits unphysical values larger than one which has to be assigned to the restricted number of loops performed.
To check that the breaking of the symmetry is an effect of the limited number of loops, the momentum distribution for $U = 0$ is studied, derived in calculations with different numbers of loops (see Fig. 5.8). The results are shown for a rather large time $t = 5.01/W$ so that the effect is clearly visible. For vanishing interaction the momentum distribution should exhibit a simple jump at k_F.

5.1 Half-Filled Model

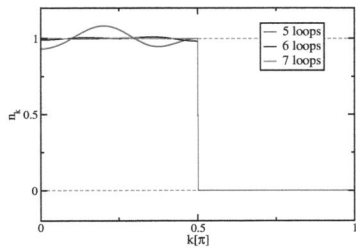

Fig. 5.8.: Momentum distribution for $U = 0$ at $t = 5.01/W$ and different numbers of loops. It can be seen that the limited number of loops leads to deviations from the analytical result for $k < k_F$.

As can be seen the results for the 5-loop and the 6-loop calculation deviate from this for $k < k_F$. The fact that the deviations appear only for $k < k_F$ is explained by the way the momentum distribution is determined. In the current approach the momentum distribution is calculated through the product $<\hat{c}_r^\dagger \hat{c}_0>$. Calculating the momentum distribution via $1- <\hat{c}_0 \hat{c}_r^\dagger>$ yields similar results, with deviations on the other side of the jump with $k > k_F$.

However, the ranges of convergence are comparable, so that the only difference is the range in momentum space where deviations are observed.

Results for the One-Dimensional Model

5.2. Influence of Doping

In the following the influence of doping on the time evolution of the model is studied. The commutators are calculated in the same manner as for the half-filled case with a Fermi vector k_F determining the filling factor. Throughout this thesis the filling factor n is to be understood as the filling for one spin species, so that the half-filled case is denoted by $n = 0.5$. Under the influence of doping the range of convergence is increased. The local expectation value used to determine the range of convergence is shown in Fig. 5.9 for various values of the filling n. The results are obtained in a 7-loop calculations, so that these are not the maximally possible results but they already show how the convergence is increased upon doping. It can be seen that the expectation value for $n = 0.1$ coincides with its initial value up to $t \approx 5.0/W$. In contrast to this the expectation value of the half-filled model strongly deviates at about $t \approx 2.0/W$. Thus on passing from $n = 0.01$ to $n = 0.5$ the convergence is decreased.

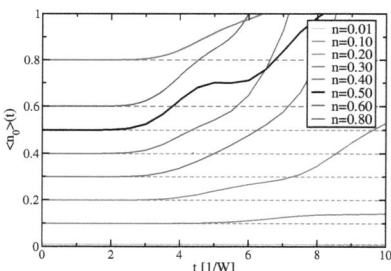

Fig. 5.9.: Local expectation value for various filling factors n for a quench to $U = 1.0W$. The results are obtained in 7-loop calculations.

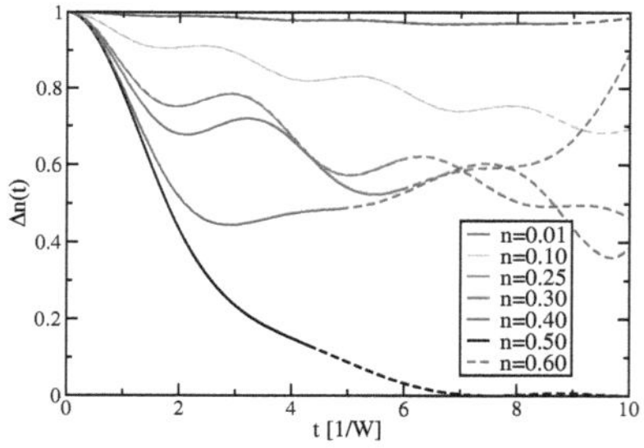

Fig. 5.10.: Jump $\Delta n(t)$ for a quench to $U = 1.0W$ and various values of the filling factor for one species n in dependence on the time t. The curve representing $n = 0.6$ lies on top of the $n = 0.4$-curve.

For values larger than 0.5 the range of convergence is increased again. Consequently the ranges of convergence for the half-filled

5.2 Influence of Doping

model are the smallest. This can be observed in the behavior of the thick black line in Fig. 5.9. For very large or very small fillings, e.g. $n = 0.01$, a much larger range of convergence is obtained.

The fact, that the deviations are not symmetric concerning particle-hole exchange, is explained by the way the time dependence of the local expectation value is determined. In the current approach the time dependence of the creation operator $\hat{c}^\dagger(t)$ is determined by the iterated equation of motion approach and the corresponding time dependence for the annihilation operator is deduced from $\hat{c}^\dagger(t)$.

The jump for a quench to $U = 1.0W$ and various fillings is shown in Fig. 5.10.

For general values of the filling n the operator $\hat{c}^\dagger_{0,\uparrow}(t)$ derived in two loops can be represented by the monomials

$$\hat{c}^\dagger_{0,\uparrow}(t) = h_0(0,t)\hat{c}^\dagger_{0,\uparrow} + h_0(-1,t)\hat{c}^\dagger_{-1,\uparrow} + h_0(1,t)\hat{c}^\dagger_{1,\uparrow} + h_1(0,0,0,t) : \hat{c}^\dagger_{0,\uparrow}\hat{c}^\dagger_{0,\downarrow}\hat{c}_{0,\downarrow} : \tag{5.3}$$

with the corresponding differential equations

$$\partial_t h_0(0,t) = -Jih_0(-1,t) - Jih_0(1,t) + Un(1-n)ih_1(0,0,0,t), \tag{5.4a}$$
$$\partial_t h_0(-1,t) = -Jih_0(0,t), \tag{5.4b}$$
$$\partial_t h_0(1,t) = -Jih_0(0,t), \tag{5.4c}$$
$$\partial_t h_1(0,0,0,t) = Uih_0(0,t) + (1-2n)Uih_1(0,0,0,t). \tag{5.4d}$$

Up to second order in t these are satisfied by the expansions

$$h_0(0,t) = 1 - (J^2 + \frac{U^2}{2}n(1-n))t^2 + O(t^3) \tag{5.5a}$$
$$h_0(-1,t) = -Jit + O(t^3) \tag{5.5b}$$
$$h_0(1,t) = h_0(-1,t) \tag{5.5c}$$
$$h_1(0,0,0,t) = Uit - \frac{U}{2}(1-2n)t^2 + O(t^3). \tag{5.5d}$$

Results for the One-Dimensional Model

valid up to second order in t. With these expansions the jump shows a quadratic correction

$$\Delta n(t) = 1 - U^2 n(1-n) t^2 + \ldots \quad (5.6)$$

depending on the filling factor for one spin species n. The quadratic term suggests that the jump is strongest reduced at half-filling ($n = 0.5$). This is shown in Fig. 5.10, where the jump for a quench to $U = 1.0W$ is depicted for various values of the filling n. The dashed curves depicted in this figure represent results derived for times which are beyond the range of convergence. Indeed the black curve representing half-filling shows the strongest decrease. The corresponding curve exhibits a shoulder. For other values of the filling the curves develop oscillations with a gradually decreased jump. Furthermore the figure shows that the curve for $n = 0.4$ and the one for $n = 0.6$ coincide, as expected due to particle-hole symmetry. This can be generalized to arbitrary fillings $\tilde{n}$ and their counterpart $1 - \tilde{n}$.

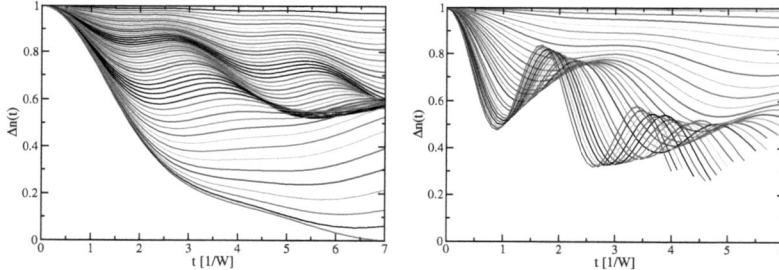

Fig. 5.11.: Behavior of the jump for $U = 1.0W$ and fillings from $n = 0.01$ to $n = 0.50$ in 0.01 steps (from top to bottom for small t).

Fig. 5.12.: Jump for the quarter-filled Hubbard model for various interaction strengths. U is increased from $0.1W$ to $3.0W$ (from top to bottom for small t).

A more detailed view of the behavior for various fillings can be found in Fig. 5.11. The jump is shown for values from $n = 0.01$ to $n = 0.50$. From $n = 0.50$ to smaller values the jump is changed

5.2 Influence of Doping

gradually from a curve with shoulder to a curve with stronger oscillations at intermediate values of the filling. For even smaller fillings the oscillations become weaker again, because the changes in the jump generally become weaker.

As exemplary result for the dependence of the jump on the interaction U under the influence of doping, results for the quarter-filled model are shown in Fig. 5.12. The interaction strength U is increased from $U = 0.1W$ to $U = 3.0W$ in steps of $0.1W$.

On increasing U the oscillations become stronger and the period of the oscillations is changed, so that the curves appear to be squeezed for increasing U.

For stronger quenches showing pronounced oscillations in the half-filled case the jump behaves similar to the curves for $U = 1.0W$ (see Fig. 5.13).

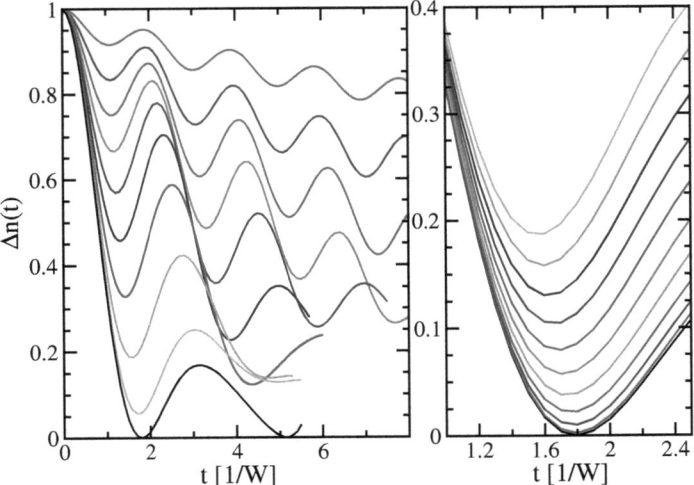

Fig. 5.13.: Jump for a quench to $U = 2.0W$ for various fillings n. Left panel: Fillings $n = 0.1$ to $n = 0.5$ in steps of 0.05 from top to bottom. Right panel: Fillings from $n = 0.40$ to $n = 0.50$ in steps of 0.01 from top to bottom.

Results for the One-Dimensional Model

In this figure results for a quench to $U = 2.0W$ are depicted. In the right panel of this figure results for smaller dopings from $n = 0.40$ to $n = 0.50$ are represented in steps of 0.01. Obviously even smallest values of the doping lead to a shift of the minima from $\Delta n = 0$ to finite values. Away from half-filling the curves show oscillations with a clear decrease underlying these oscillations in time.

To demonstrate that the appearance of oscillations showing zeros is not just shifted to higher values of the interaction U upon doping, extremely strong quenches to $U = 100W$ are discussed. The resulting jump for various fillings is shown in Fig. 5.14. The curves show coherent oscillations with a fixed period (see Sect. 5.4.1). For half-filling the curve oscillates between zero and values close to unity. Upon doping the minima are shifted to higher values, as depicted in Fig. 5.15. In this figure the minima Δn_{min} are given in dependence on the filling factor n with $\Delta n_{min} = 0$ for $n = 0.50$ and $\Delta n_{min} \to 1$ for $n \to 0$.

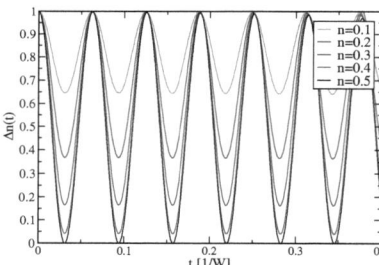

Fig. 5.14.: Jump $\Delta n(t)$ for a large quench to $U = 100W$ for various fillings n. The curves show oscillations with the minima shifted upwards under the influence of doping.

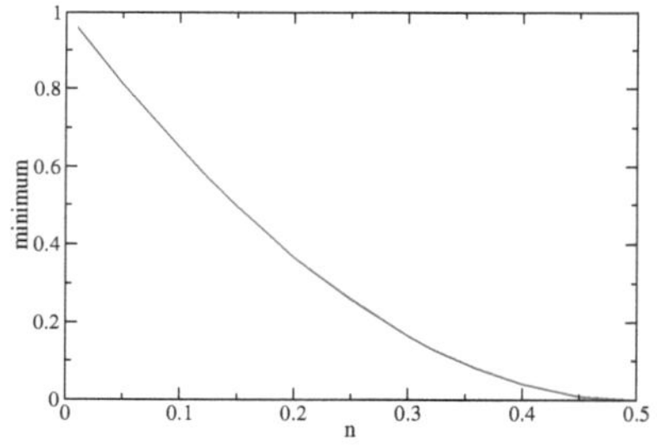

Fig. 5.15.: Value of the minima observed in quenches to $U = 100W$ in dependence on the filling factor n for one spin species.

The upwards shift of the minima is explained by the Fourier transform of the one-particle terms $h_k(t)$ used to calculate the jump $\Delta n(t) = |h_k(t)|^2_{k \in FS}$. For half-filling $h_k(t)$ is real, whereas it displays a non-vanishing imaginary part for various fillings (see

5.2 Influence of Doping

Fig. 5.16). The non-vanishing imaginary part leads to the shift of the minima upwards under the influence of doping. The 2-loop calculation already indicates the qualitatively different behavior for half-filling and under the influence of doping. For half-filling the expansions in time given in Eqs. 5.5a to 5.5d reveal prefactors which have either a vanishing real or a vanishing imaginary part. Upon doping the prefactor for the three-particle term reads $h_1(0,0,0,t) = Uit - \frac{U}{2}(1-2n)t^2 + O(t^3)$. It contains real and imaginary parts.

At half-filling the zeros in the oscillations of the jump for strong quenches are attributed to collapse and revival behavior of the momentum distribution. A similar behavior is observed experimentally for bosons in the matter wave of Bose-Einstein condensates [8].

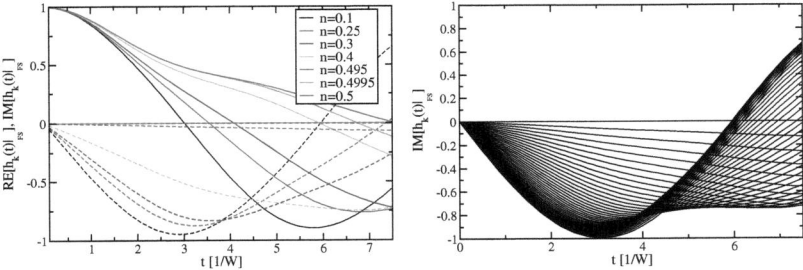

Fig. 5.16.: Real(solid lines) and imaginary part(dashed lines) of the Fourier transformed prefactor of the one-particle terms $h_k(t)$ for various fillings n and $U = 1.0W$. For half-filling the imaginary part vanishes.

Fig. 5.17.: Imaginary part of $h_k(t)$ for $U = 1.0W$. Upon doping the imaginary part is increased. The curves are shown for $n = 0.5$ to $n = 0.01$ in steps of 0.01 (from top to bottom for small t).

Due to the non-vanishing imaginary part the jump is shifted upwards. Thus upon doping the behavior of the jump changes qualitatively. This change is dicussed further in Sect. 5.4.2.

Upon doping the jump is not only reduced but also shifted according to the Fermi vector k_F.

Results for the One-Dimensional Model

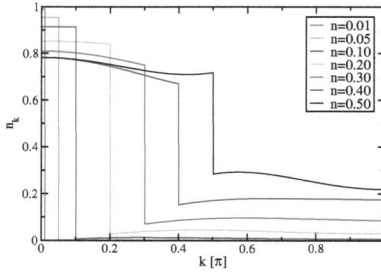 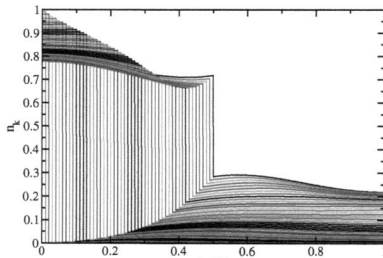

Fig. 5.18.: Full momentum distribution for a quench to $U = 1.0W$ and various fillings.

Fig. 5.19.: Momentum distribution for values of n between $n = 0.5$ and $n = 0.01$ in steps of 0.01.

The momentum distribution for a jump to $U = 1.0W$ and various fillings is shown in Figs. 5.18 and 5.19. The momentum distribution is given for a fixed time $t = 2.0/W$, where the half-filled model exhibits a momentum distribution which is first decreased and then slightly increased close to the Fermi surface.

Under the influence of doping the non-monotonic behavior is lost and the momentum distribution for $n = 0.4$ shows a pure decrease for $k < k_F$ followed by an increase on the other side of the discontinuity. Doping the system further away from half-filling the increase becomes weaker until a nearly flat distribution is reached on both sides of the jump. In Fig. 5.19 it is shown how the curvature in the momentum distribution is gradually lost. The momentum distribution is depicted for fillings from $n = 0.50$ to $n = 0.01$ in steps of 0.01.

To compare this behavior to the behavior of the momentum distribution for different fillings the time evolution of the momentum distribution for quenches to $U = 1.0W$ and filling factors $n = 0.1$ and $n = 0.4$ is shown in Figs. 5.20 and 5.21.

5.2 Influence of Doping

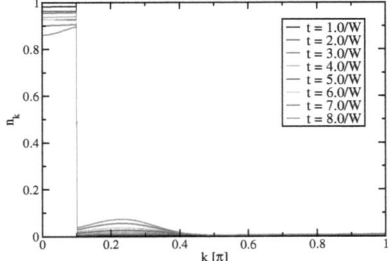
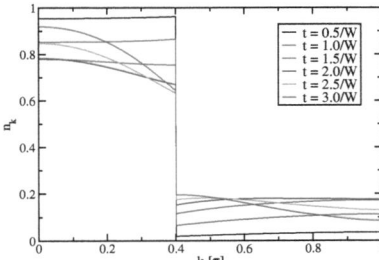

Fig. 5.20.: Momentum distribution for a quench with $U = 1.0W$ and a filling factor $n = 0.1$.

Fig. 5.21.: Corresponding results for a filling factor of $n = 0.4$ and various times.

The shift in the momentum distribution according to the filling factor can be seen. Comparing both cases the momentum distribution for $n = 0.1$ is much flatter than the one for $n = 0.4$. As explained the jump is only gradually decreased for small n so that the momentum distribution is much flatter for $n = 0.1$ even though much larger times are considered.

Results for the One-Dimensional Model

5.3. Bosonization

One-dimensional models without or with small gaps play a special role in condensed matter theory as these are accessible to bosonization techniques. This section deals with the predictions gained from bosonization theory and the question in how far the time evolution of the system can be understood by the use of these results. It turns out that the results for the jump can indeed be described by a power law decay. But the exponents differ from the usual bosonization results.

The one-dimensional Hubbard model is integrable, which implies the existence of an infinite number of conserved quantities. Thus the integrability is assumed to influence the dynamics of the model strongly. Due to the conserved quantities the system retains memory of its initial state. Consequently the model does not access the full energy surface [23, 28]. One-dimensional models are accessible to many analytical and numerical methods in equilibrium. One powerful method for one-dimensional models are bosonic field theories [85–91]. Bosonic field theories can be applied to gapless one-dimensional models which exhibit a linear dispersion at low energies. In these approaches the bosonic fields are often assumed to be interaction-free. An example of the application are spinless fermions. In models comprising spin the leading bosonic interaction is captured by the sine-Gordon model [89, 91].

At present it is still unclear if renormalization group approaches are applicable to systems far from equilibrium, because it is unclear whether processes which are neglected in renormalization group approaches matter for systems out of equilibrium [92–94]. The dynamics after the quench is governed by high-energy states which are usually not captured by these techniques [95]. Thus it is still an open question if the same or similar field theories can be applied to systems out of equilibrium [96–98].

There are several studies of quenches in bosonic models so that the dynamics of these systems is well-understood by now [36, 37, 72, 93]. For interaction-free bosonic quenches the time evolution can be computed exactly. Also the sine-Gordon model is studied in this context [38, 60, 92, 99–101]. For spinless fermions Karrasch et al.

5.3 Bosonization

claim that the dynamics is described by bosonic field theories [96]. It is suggested that the decay of the jump shows a power law behavior with exponents given by Bethe ansatz results. Besides, a continuous Bose gas did not show thermalization [102]. For a lattice Bose Hubbard model a dependence of the dynamics on the initial state was observed. From these different observations the question arises how generic these findings are for a one-dimensional model exposed to a quench.

In a first step a quench in the spinless fermion model is studied as testbed for the description with bosonization theories.

5.3.1. Spinless Fermion Model

The first model under study is an integrable model of spinless fermions. The corresponding Hamiltonian is presented by

$$H_{NN} = -J \sum_{\langle i,j \rangle} \left(\hat{c}_i^\dagger \hat{c}_j + \text{h.c.} \right) + U(t) \sum_i \hat{n}_i \hat{n}_{i+1} \tag{5.7}$$

with the hopping element J and a nearest-neighbor repulsion (NN) with interaction strength U. Again $\hat{c}_i^\dagger$ creates a particle at site i and the operator $\hat{n}_i = \hat{c}_i^\dagger \hat{c}_i$ counts the number of particles at site i. This model can be mapped by a Jordan-Wigner transformation [103] to an anisotropic spin $S = \frac{1}{2}$ XXZ-chain [104]. As integrable model it is solvable by Bethe ansatz [105–108].

Results for the jump in the spinless fermion model with nearest neighbor interaction are depicted in Fig. 5.22. The results are shown for a quench with $U = 0.25W$ and different numbers of loops m. The jump starts again at $\Delta n(0) = 1$ and is then decreased with time t. It can be seen that the convergence is quickly increased on increasing m.

A comparison of results for various values of the nearest-neighbor repulsion is shown in Fig. 5.23. These results are obtained by an 11-loop calculation in the iterated equation of motion approach.

Results for the One-Dimensional Model

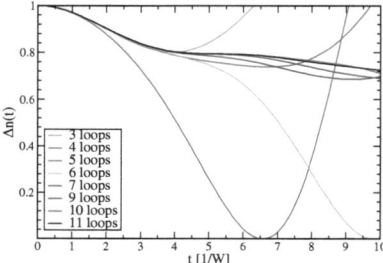
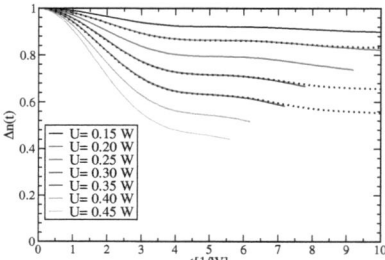

Fig. 5.22.: Jump $\Delta n(t)$ in the half-filled spinless fermion model. The results are shown for an interaction strength of $U = 0.25W$ and various numbers of loops. The convergence is clearly increased on increasing loopnumber.

Fig. 5.23.: Results for the jump $\Delta n(t)$ for increasing values of the nearest-neighbor repulsion U (from top to bottom) obtained in a calculation with 11 loops. The symbols display exemplary results obtained by infinite-size DMRG [96].

Additionally this figure shows exemplary results obtained by Karrasch et al. [96] by the use of a time dependent variant of infinite-size DMRG [52, 109, 110]. In the range where the equation of motion results are converged the data agree excellently.

When comparing the range of convergence for the two approaches the fact must be considered that the computer code used in the iterated equation of motion approach is optimized for the description of spinful models, whereas the DMRG program is adapted to the spinless fermions model in one dimension. Besides, the equation of motion approach is designed to also describe the dynamics of two-dimensional models. It allows to study arbitrarily large interaction strengths and becomes exact in the limit of infinite interactions.

For the spinless fermions the computational resources allow to perform 11 loops just like for the Hubbard model. An exemplary 11-loop calculation for the half-filled spinless fermion model contains 374573 monomials and 32298856 terms on the right hand side of the differential equations. In order to quantify the range of

5.3 Bosonization

convergence for this model results for different numbers of loops are compared to the 11-loop calculation. The absolute difference of an m-loop calculation from the 11-loop calculation for a quench to $U = 0.25W$ is shown in Fig. 5.24 in a double logarithmic plot. It can be seen how each loop increases the range of convergence.

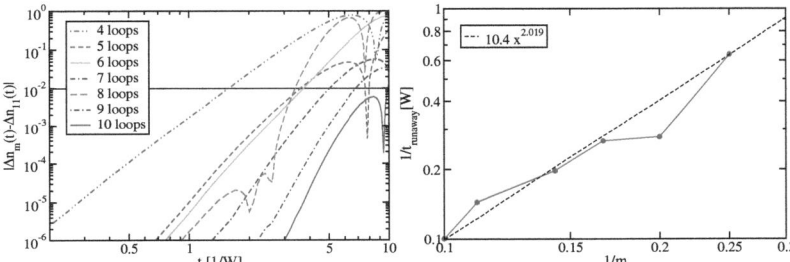

Fig. 5.24.: Absolute difference of the jump for various numbers of loops m relative to the 11-loop result for a quench to $U = 0.25W$. The results are shown in a double logarithmic plot. The horizontal black line indicates a threshold of 0.01 used to determine the runaway times.

Fig. 5.25.: Double logarithmic plot of the inverse runaway time versus the inverse number of loops m. The dashed line represents a linear fit. The fit indicates a dependence of the inverse runaway time on the inverse loop number with an exponent of about 2.02.

The horizontal black line indicates the threshold of 0.01 used for the determination of the runaway time. As explained in Sect. 3.5 the choice of the threshold is arbitrary. Thus a value of 0.01 is chosen to visualize the increase of the runaway times with increasing loop number m. The resulting inverse runaway times are depicted in Fig. 5.25.

For a threshold of 0.01 the inverse runaway times are described by a power law with an exponent of about 2.02. A different choice for the threshold results in a slightly varying exponent around 2 and a different prefactor for the linear fit. Again the exponent and hence the dependence on the time t is even better than predicted by the approach (see Sect. 3.2.2). Also for this model the increase of the

Results for the One-Dimensional Model

runaway time on increasing loop number is superlinear, just like it was observed in the Hubbard model as explained in Sect. 3.5.

5.3 Bosonization

5.3.2. General Concepts

At the beginning of this section the results of bosonization are briefly recalled. The one-dimensional Hubbard model and the one-dimensional spinless fermions model studied in the present thesis are Luttinger liquids in the sense that the low energy physics of these models in equilibrium can be described by the Tomonaga-Luttinger model [89, 91, 111, 112] with renormalized coupling constants.

For a one-dimensional model of correlated particles the Luttinger model is the low energy fixed point of the model if a gapless branch in the dispersion exists. Under these conditions the Tomonaga-Luttinger model describing interaction-free bosons can be used to describe the low-energy physics of other models [86, 105, 106, 113]. For the single-particle density in the Tomonaga-Luttinger model a power law behavior was observed [87, 114]. The jump in the momentum distribution of the Tomonaga-Luttinger model after a quench has been studied [36, 37, 72, 93] and Cazalilla found a power law

$$\Delta n(t) = \left(\frac{R_0}{2vt}\right)^{\alpha^2} \tag{5.8}$$

with the dressed velocity v [37]. The exponent α is given by the equilibrium exponent 2γ via $\alpha^2 = 4\gamma(\gamma+1)$. To apply Eq. 5.8 the equilibrium exponent γ has to be calculated. In the Tomonaga-Luttinger model γ is given by the parameter Θ of the Bogoliubov-transformation via $\gamma = \sinh^2(\Theta)$ [72].

In 2009 Uhrig extendet the power law according to

$$\Delta n(t) = \left[\frac{r^2}{r^2 + (2vt)^2}\right]^{2\gamma(\gamma+1)} \tag{5.9}$$

for a spinless model described by the Tomonaga-Luttinger model for low energies, introducing the characteristic length scale of the interaction r [72]. This power law agrees with the one derived by Cazalilla in the limit $t \to \infty$. In Eq. 5.9 the interaction is assumed

Results for the One-Dimensional Model

to range over all momenta from $-\infty$ to ∞. A finite range in momentum space results in oscillations in the dynamics. The occurrence of oscillations due to high energy cutoffs is a generic feature of the quench dynamics [37, 60, 74, 93]. However such a finite momentum range is inherent to microscopic models. In these models the Brillouin zone is finite which leads to a finite range of the interaction in momentum space.

To determine γ in the spinless fermions model bosonization techniques [115, 116] can be applied. The exponent γ is given by the standard bosonization parameter K through the definition

$$\gamma = \frac{K + \frac{1}{K} - 2}{4}. \tag{5.10}$$

The exponent is completely determined by the anomalous dimension K.

Thus γ is the same as in equilibrium, but the exponent 2γ is replaced by $2\gamma(\gamma+1)$ [72, 87, 117]. This replacement is inherent to the Tomonaga-Luttinger model and not dependent on the microscopic model under study. The replacement leads to a factor of 2 for small values of γ. A similar effect has been observed by Moeckel and Kehrein [47] for an infinite-dimensional model. They addressed the time-dependence of the infinite dimensional model by a second order in U approach.

5.3 Bosonization

5.3.3. Results for Spinless Fermions

5.3.3.1. Exponents

Even though the explicit formula for $\Delta n(t)$ is given, the anomalous dimension K and the dressed velocity v still have to be determined. In the first step the values expected from bosonization of the spinless fermion model in equilibrium are used [87, 89, 118, 119]. Since the model is considered at zero temperature the equilibrium refers to the ground state and its immediate vicinity, i.e., its elementary excitations. Consequently the Bethe ansatz can be used to determine K and v [85, 107, 108]. The exact solution for the anomalous dimension reads [84, 86, 90]

$$K_{\text{GS}} = \frac{\pi}{2(\pi - \arccos(2U/W))}, \quad (5.11)$$

and the velocity is given by [120, 121]

$$v_{\text{GS}} = \frac{\pi \sin(\arccos(2U/W))}{2\arccos(2U/W)}. \quad (5.12)$$

As these values correspond to the behavior of the model at lowest energies which means in the vicinity of the ground state, they are labelled by the subscript 'GS'.

The resulting values for the anomalous dimension K are depicted in Fig. 5.26.

From the anomalous dimension K the equilibrium exponent γ can be calculated. For the GS results the equilibrium exponent γ is derived to be

$$\gamma = \frac{\arcsin((2U/W)^2)}{\pi^2 + 2\pi \arcsin(2U/W)}. \quad (5.13)$$

Results for the One-Dimensional Model

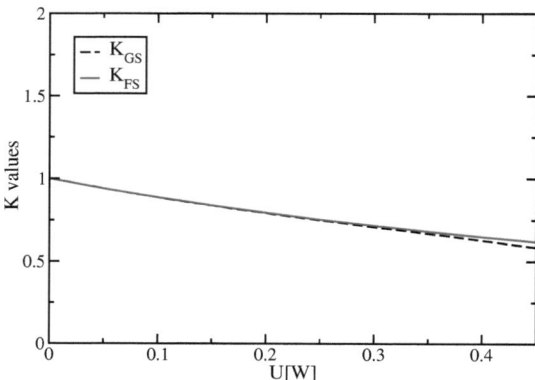

Fig. 5.26.: Anomalous dimension K obtained by bosonization. Dashed lines: Results from bosonization around the groundstate (GS) derived by Bethe ansatz techniques. Solid lines: Results from bosonization around the initial state (FS). In this approach Fermi seas are used as initial states.

Using the calculated exponents the decay of the jump can be fitted to the data derived by the equation of motion approach. The cutoff length r of the interaction is used as fit parameter. It takes values from 0.2 to 0.6 on increasing values of the interaction strength U. The resulting curves are shown as dashed black lines in Fig. 5.27. The corresponding curves for the iterated equation of motion result are given as coloured solid lines. These are derived from 11-loop calculations. As can be seen the fits describe the decay of the jump for the respective interaction strength U appropriately. The only feature missed by the fits are the oscillations lying on top of the decay. As explained these oscillations are to be attributed to the momentum cutoff of the interaction in microscopic models [60, 74, 93]. This agrees with the work by Karrasch et al. [96], where a description of the jump by the GS exponents was concluded.

Inspecting Fig. 5.27 it can be seen that the agreement between

5.3 Bosonization

the numerical equation of motion curve and the fitted bosonization curve deteriorates for larger values of the interaction U. For instance, the curves for $U = 0.45W$ and $U = 0.5W$ show differences between the curve and the fit.

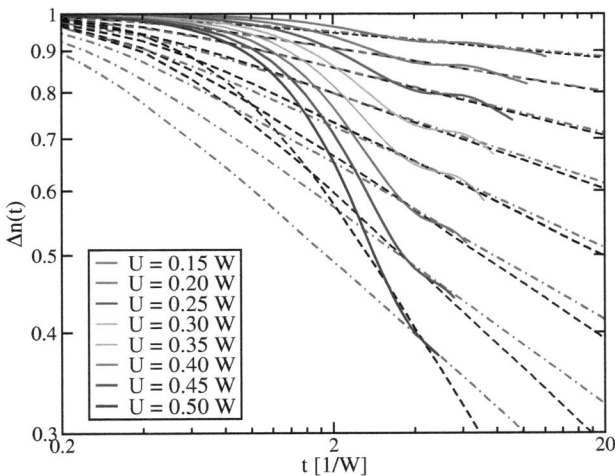

Fig. 5.27.: Coloured solid lines: Results for the jump $\Delta n(t)$ for various U as derived by the iterated equation of motion approach in 11 loops. Dashed black lines: Fitted curve for the GS exponents as given by Eq. 5.11. Dashed dotted red lines: Fits for the FS exponents given in Eq. 5.14a.

Thus the question arises whether a different exponent describes the time dependence of the jump more properly. Although the GS exponents capture the dynamics of the Tomonaga-Luttinger model describing the physics near the ground state it is possible that microscopic models like the ones considered in this thesis or models with boson-boson interactions require different exponents. In such models the behavior near the ground state may in general be different from the behavior at higher energies. Thus a field theoretic description of such microscopic models after interaction quenches, where highly excited states are reached, is expected to depend on the initial conditions and on the strength of the quench. An example for such a dependence is found in the sine-Gordon model [92, 101]. This model is quenched in a different protocol, where the sine term is switched on only on adiabatically long times after the first sudden quench. A renormalization group approach applied to this model revealed an explicit dependence of the anomalous dimension on the initial conditions.

In the next step an alternative calculation of K and v is discussed, which accounts for a dependence of the parameters on the initial state. The initial state considered in this thesis is the non-interacting Fermi sea. On the time scales considered here, i.e., short and intermediate times after the quench, the dynamics is governed by the iterated gradual excitation of particle-hole pairs [72]. Consequently only a limited number of particle-hole pairs has to be considered for the description of the system on moderate times. Thus bosonization of the density fluctuations around the initial state suggests itself. As the initial state is the Fermi sea, the corresponding parameters are labelled by the subscript 'FS' in the following. Such a bosonization around the Fermi sea corresponds to a bosonization in leading order in U since no feedback effects due to the interaction need to be included.

To calculate the bosonization results in leading order in U the g-ology notation [122, 123] is used. The name g-ology stems from the fact that the coupling constants are denoted by g. In this notation the anomalous dimension and the renormalized velocities are given via [116, 122]

$$K_{\text{FS},\nu} = \sqrt{\frac{\pi v_F + g_{4,\nu} - g_{2,\nu}}{\pi v_F + g_{4,\nu} + g_{2,\nu}}} \qquad (5.14\text{a})$$

and [124]

$$v_{\text{FS},\nu} = \sqrt{\left(v_F + \frac{g_{4,\nu}}{\pi}\right)^2 - \left(\frac{g_{2,\nu}}{\pi}\right)^2} \qquad (5.14\text{b})$$

where ν denotes the different channels in the dispersion. For a spin-

5.3 Bosonization

ful model these read $v \in \sigma, \rho$ for the spin (σ) and the charge channel (ρ). The parameters K_v and v_v are sufficient to describe the low-energy sector of each degree of freedom v. In the case of spinless fermions only one channel exists, so that the subscript can be omitted. Then Eqs. 5.14a and 5.14b simplify to

$$K = \sqrt{\frac{\pi v_F + g_4 - g_2}{\pi v_F + g_4 + g_2}} \tag{5.15}$$

$$v = \sqrt{\left(v_F + \frac{g_4}{\pi}\right)^2 - \left(\frac{g_2}{\pi}\right)^2}. \tag{5.16}$$

In this formula g_2 and g_4 denote forward scattering (with momentum transfer $\ll k_F$) between particles on different branches and on the same branch of the linearized dispersion [123]. In leading order in U the g's read [90]

$$g_2 = 4U \tag{5.17a}$$
$$g_4 = 2U \tag{5.17b}$$

resulting in [85, 86, 88, 89]

$$K_{FS} = \sqrt{(\pi v_F - 2U)/(\pi v_F + 6U)} \tag{5.18}$$

$$v_{FS} = \frac{1}{\pi}\sqrt{(\pi v_F + 2U)^2 - 8U^2}. \tag{5.19}$$

Expanding both results for the anomalous dimensions K into a Taylor series yields

$$K_{GS} = \frac{\pi}{2(\pi - \arccos(\frac{2U}{W}))} \tag{5.20a}$$

$$\approx 1 - \frac{2}{\pi}2U + \frac{4}{\pi^2}(2U)^2 - \left(\frac{8}{\pi^3} - \frac{1}{3\pi}\right)(2U)^3 + \dots \tag{5.20b}$$

Results for the One-Dimensional Model

and

$$K_{FS} = \sqrt{\frac{1-\frac{2U}{\pi}}{1+\frac{6U}{\pi}}} \tag{5.21a}$$

$$\approx 1 - \frac{2}{\pi}2U + \frac{4}{\pi^2}(2U)^2 - \frac{10}{\pi^3}(2U)^3 + ... \tag{5.21b}$$

with $k_F = \frac{\pi}{2}$ for the half-filled model. Obviously the formulae agree up to second order in U. The values for the anomalous dimension derived by bosonization around the Fermi sea are included in Fig. 5.26 as solid red lines. With the anomalous dimension K_{FS} and the velocity v_{FS} the power law given in Eq. 5.9 can be fitted to the numerical data. The resulting curves are depicted as red dashed dotted lines in Fig. 5.27. The fitted interaction range r takes values from 0.2 to 0.1 on increasing U. As can already be seen from the anomalous dimension K in Fig. 5.26 the differences in the results of the FS and the GS bosonization are fairly small. However, for larger values of the interaction the FS exponents describe the numerical data better. Of course, the ranges of convergence are much smaller for larger U but it can still be seen, how the curves for $U = 0.45W$ and $U = 0.5W$ fit to the FS exponents while deviating from the GS curves. The bosonization around the Fermi sea is restricted to values of U below $\pi v_F/2 \approx 1.57W$. At this point K_{FS} and v_{FS} vanish. This unphysical behavior is due to the breakdown of this type of bosonization for too large U. For large values of U additional effects like the curvature of the dispersion [105, 106] have to be considered, which may spoil the result. Besides, umklapp scattering, which appears in higher orders of U, may become important. Thus a breakdown of the FS description for large U is not surprising.

The results presented suggest that the dynamics can indeed be described by bosonization but the exponents differ from the ones expected in equilibrium.

5.3 Bosonization

5.3.3.2. Quench Energy

To measure how far a system is driven from the equilibrium by the quench, the quench energy ΔE of a microscopic model is defined as

$$\Delta E := \langle \text{FS}|\hat{H}(t>0)|\text{FS}\rangle - \langle \text{GS}|\hat{H}(t>0)|\text{GS}\rangle \quad (5.22)$$

with $\langle \text{FS}|$ denoting the Fermi sea and $\langle \text{GS}|$ the ground state of the quenched Hamiltonian. As the Fermi sea is the initial state in the quench protocol the quench energy represents the excitation energy of the system above its ground state that is induced by the quench. Since energy is conserved in a closed quantum system ΔE is also conserved. This is in contrast to quenches in imaginary time. For quenches in imaginary time the time evolved state reads $|t\rangle = e^{-H\tau}|\text{FS}\rangle$. Consequently energy is no longer conserved [96].

For the first part $\langle \text{FS}|\hat{H}(t>0)|\text{FS}\rangle$ the expectation value of the quenched Hamiltonian with respect to the Fermi sea has to be calculated. It can be determined through normal ordering using Wick's theorem

$$\left(\hat{n}_i - \frac{1}{2}\right)\left(\hat{n}_{i+1} - \frac{1}{2}\right) = \hat{n}_i\hat{n}_{i+1} - \frac{1}{2}\hat{n}_i - \frac{1}{2}\hat{n}_{i+1} + \frac{1}{4}. \quad (5.23)$$

As normal ordered terms do not contribute in the expectation value, this simplifies to

$$\begin{aligned}\left\langle\left(\hat{n}_i - \frac{1}{2}\right)\left(\hat{n}_{i+1} - \frac{1}{2}\right)\right\rangle &= \langle \hat{c}_{i+1}^\dagger \hat{c}_{i+1}\rangle\langle \hat{c}_i^\dagger \hat{c}_i\rangle \\ &+ \langle \hat{c}_i^\dagger \hat{c}_{i+1}\rangle\langle \hat{c}_i \hat{c}_{i+1}^\dagger\rangle \\ &- \frac{1}{2}\langle \hat{c}_i^\dagger \hat{c}_i\rangle - \frac{1}{2}\langle \hat{c}_{i+1}^\dagger \hat{c}_{i+1}\rangle + \frac{1}{4}.\end{aligned} \quad (5.24)$$

Results for the One-Dimensional Model

At half-filling the expectation values read

$$\langle \hat{c}_i^\dagger \hat{c}_i \rangle = \frac{1}{2} \tag{5.25a}$$

$$\langle \hat{c}_i^\dagger \hat{c}_{i+1} \rangle = \frac{1}{\pi} \tag{5.25b}$$

$$\langle \hat{c}_i \hat{c}_{i+1}^\dagger \rangle = -\frac{1}{\pi} \tag{5.25c}$$

which implies

$$\langle \left(\hat{n}_i - \frac{1}{2}\right)\left(\hat{n}_{i+1} - \frac{1}{2}\right) \rangle = -\frac{1}{\pi^2} \tag{5.26}$$

for the interaction and

$$-\langle \left(\hat{c}_i^\dagger \hat{c}_{i+1} + h.c.\right) \rangle = -\frac{2}{\pi} \tag{5.27}$$

for the kinetics. To calculate the ground state energy of the quenched Hamiltonian $\langle GS|\hat{H}(t>0)|GS\rangle$ it is convenient to use a Jordan-Wigner transformation and map the spinless fermion model to an XXZ-chain [125]

$$\hat{H}_{XXZ} = -\frac{1}{2} \sum_i \left(\sigma_i^x \sigma_{i+1}^x + \sigma_i^y \sigma_{i+1}^y - \Delta \sigma_i^z \sigma_{i+1}^z + h.c. \right) \tag{5.28}$$

with the anisotropic coupling Δ. For $\Delta < 1.0$ the system is metallic.

The ground state energy per lattice site of the XXZ-chain can be determined exactly to be [108, 126, 127]

$$e_0 = \frac{\Delta}{4} - \sin(\arccos(\Delta))^2 \int_{-\infty}^{\infty} \frac{dx}{2\cosh(\pi x)(\cosh(2x\arccos(\Delta)) - \Delta)}. \tag{5.29}$$

5.3 Bosonization

Expanding the ground state energy yields a linear part

$$e_{0,lin} = -\frac{\Delta}{\pi^2} - \frac{1}{\pi} \qquad (5.30)$$

which equals the term found by normal ordering of the Hartree and the Fock term. The quenched energy in dependence on the interaction U can be found in Fig. 5.28.

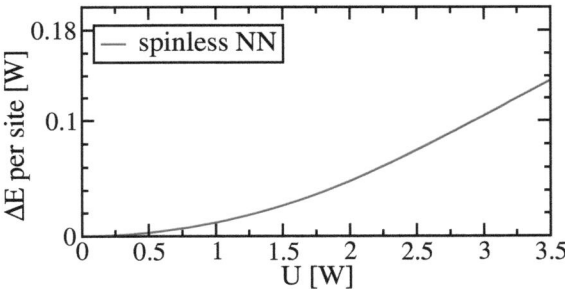

Fig. 5.28.: Quench energy ΔE per site (see Eq. 5.22) for the half-filled spinless fermions model in dependence on the interaction strength U.

It can be seen that the excitation energies stay fairly small $< 3 \cdot 10^{-3} W$ for quenches below the phase transition at $U = 0.5W$. The small quench energy explains why the GS exponents and the FS exponents coincide and both yield a good description of the decay of the jump. For larger quenches the excitation energy is increased and so is the difference between a description with the GS and the FS exponents. This explains why only at large U the difference between the anomalous dimensions K_{GS} and K_{FS} becomes discernible with the effect that the FS exponents fit the slope better (see for instance $U = 0.45W$).

5.3.4. Bosonization in the Quarter-Filled Hubbard Model

5.3.4.1. Exponents

In this section bosonization for the quarter-filled Hubbard model is discussed. This model exhibits sizeable differences in the values of the anomalous dimension derived by bosonization around the ground state (GS) and bosonization around the Fermi sea (FS). Thus the role of these parameters can be discussed further.

To stay within the metallic phase weaker quenches ($U \lessgtr 1.11W$) are considered in this context. A discussion of strong quenches is given in Sect. 5.4.1. Quarter-filling is chosen to suppress umklapp scattering at least to leading order. At half-filling these processes occur [86, 91] as an additional term and the sine-Gordon model is recovered. The additional term leads to an insulating state [91, 128, 129]. Consequently quarter-filling is chosen to avoid quenches to the Mott insulating phase.

For spinful models the Tomonaga-Luttinger Hamiltonian consists of the non-interacting sum of the charge and the spin part of the model. Thus the one-particle correlation function which determines the jump Δn is calculated as product of the responses in the two channels. Consequently the formulae for the jump are modified to

$$\Delta n(t) = \left[\frac{r_\rho^2}{r_\rho^2 + (2v_\rho t)^2}\right]^{\gamma_\rho(\gamma_\rho+1)} \left[\frac{r_\sigma^2}{r_\sigma^2 + (2v_\sigma t)^2}\right]^{\gamma_\sigma(\gamma_\sigma+1)} \qquad (5.31)$$

where the spin channel carries the subscript σ and the charge channel is denoted by ρ [72]. The fit parameters r_σ and r_ρ denote the interaction ranges in the respective channel. Just like in the case of spinless fermions the exponent of each channel is given by the corresponding equilibrium exponent γ_ν [72] via

$$\gamma_\nu(\gamma_\nu + 1). \qquad (5.32)$$

Comparing the formulae for the jump the exponent in each channel

5.3 Bosonization

takes half the value of the exponent in the spinless case just like in equilibrium [87, 89]. The results for spinless models can easily be recovered by the above formula Eq. 5.31 by setting $K_\rho = K_\sigma = K$ and $v_\sigma = v_\rho = v$ [117]. In the Hubbard model the electron-electron interaction leads to $v_\rho \neq v_\sigma \neq v_F$, inducing spin-charge separation [87, 116, 130, 131]. The parameters $K_{\nu,\text{GS}}$ and $v_{\nu,\text{GS}}$ for bosonization in the vicinity of the ground state are obtained from ground state properties following from Bethe ansatz [129, 132–134].

Since no explicit formula for the Bethe ansatz results in the Hubbard model exist, the parameters have to be determined by solving the coupled integral equations numerically. For the quarter-filled model this task is rather demanding. Here only the results for the parameters are discussed. The derivation of these parameters is explained in App. C. In this appendix the solution of the integral equations for the quarter-filled Hubbard model are explained.

In the second approach the FS exponents are derived by bosonization in the vicinity of the Fermi sea. Thus the Hubbard model is mapped to the Tomonaga-Luttinger model [135], where in addition to the terms already present in the Tomonaga-Luttinger model backscattering is also included. Backscattering is denoted by g_1 with the processes $g_{1,\perp}$ and $g_{1,\|}$. For these parameters renormalized values have to be determined by the use of renormalization group techniques. The parameters of the g-ology approach for the two channels are given by the parallel and perpendicular processes via

$$g_{i,\rho} = \frac{1}{2}(g_{i,\|} + g_{i,\perp}) \quad g_{i,\sigma} = \frac{1}{2}(g_{i,\|} - g_{i,\perp}) \tag{5.33}$$

which can be used in Eqs. 5.14a and 5.14b. Again the FS parameters coincide with the bosonization results in leading order of U [136]. The parameters satisfy $g_{i,\sigma} = -U$ and $g_{i,\rho} = U$ [88, 89, 137] leading to

$$K_{\rho,\text{FS}} = \sqrt{2\pi v_F/(2\pi v_F + 2U)} \tag{5.34a}$$

$$K_{\sigma,\text{FS}} = \sqrt{2\pi v_F/(2\pi v_F - 2U)}. \tag{5.34b}$$

Results for the One-Dimensional Model

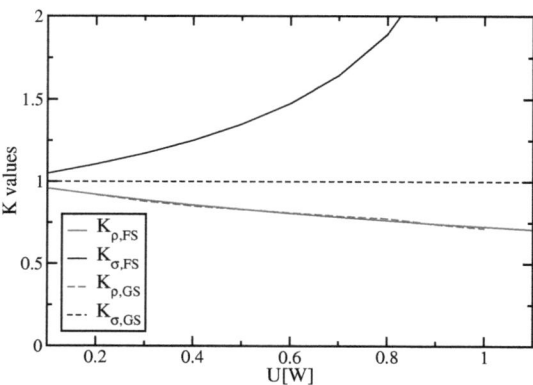

Fig. 5.29.: Anomalous dimension K_ν with $\nu \in \sigma, \rho$ obtained by bosonization. Dashed lines: Results from bosonization around the GS derived by Bethe ansatz techniques. Solid lines: Results from bosonization around the FS. The Bethe ansatz equations are difficult to evaluate so that slight inaccuracies imply some minor wiggling of the dashed curve for K_ρ for larger interaction strengths U.

The corresponding velocities read

$$v_{\rho,\text{FS}} = v_F \sqrt{1 + U/(\pi v_F)} \qquad (5.35)$$

$$v_{\sigma,\text{FS}} = v_F \sqrt{1 - U/(\pi v_F)}. \qquad (5.36)$$

The results for the anomalous dimensions are included in Fig. 5.29. It can be seen that the anomalous dimensions for the charge channel agree surprisingly well $K_{\rho,\text{FS}} \approx K_{\rho,\text{GS}}$. However, the parameters for the spin channel K_σ differ significantly. This can be understood from an analysis of the underlying sine-Gordon model [91]. Renormalization group approaches lead to an extremely slow convergence of the spin parameter $K_{\sigma,\text{GS}}$, so that its final value is reached only on exponentially small energy scales. The energy scale ϵ introduces an exponentially large distance

5.3 Bosonization

$\vec{r}$ according to $\epsilon \approx \frac{v_F}{|\vec{r}|}$ with the Fermi velocity v_F. To capture the flow of $K_{\sigma,\text{GS}}$ correctly the equal-time Green function has to be known for two points at distance $\vec{r}$. Due to the turning on of the interactions correlations are created by the quench, which spread out on distances $|\vec{r}| = v_{\text{max}} t$ after the quench with v_{max} denoting the maximal velocity of quasi-particles. Combining this with the energy scale ϵ a term proportional to $\frac{1}{t}$ is obtained. Thus $\frac{1}{t}$ acts as a low-energy cutoff. Consequently exponentially large times have to be accessed to see the effects of processes on exponentially low energies. However, there may be other cutoffs in the system due to which the flow of the parameters is stopped before lowest energies are reached. These additional cutoffs and consequences for the long time behavior are adressed in Sect. 5.3.5.

Having computed the parameters K_ν and v_ν, the decay of the jump given by Eq. 5.31 can be fitted to the data. In the GS approach $K_{\sigma,\text{GS}} = 1$ holds so that the corresponding exponent $\gamma_{\sigma,\text{GS}}$ vanishes and thereby the dependence of Δn on the range $r_{\sigma,\text{GS}}$. For the GS exponents the range $r_{\rho,\text{GS}}$ is the only fitting parameter. The resulting curves for a quench to $U = 0.8W$ in the quarter-filled Hubbard model are shown in Fig. 5.30 in a double logarithmic plot and in a linear plot in the inset of this figure.

Obviously the significant differences in the anomalous dimensions $K_{\sigma,\text{GS}}$ and $K_{\sigma,\text{FS}}$ lead to sizeable differences in the power laws. Clearly the FS exponents capture the decay of the jump much better than the GS exponents. For the FS exponents the interaction ranges are fitted to be $r_{\rho,\text{FS}} \approx 1$ and $r_{\sigma,\text{FS}} \approx 0.6$. The slope of the Δn curve derived by the equation of motion approach is captured very well. In contrast to this a fit with the GS exponents is only possible for a very limited range in time as can be seen in the inset of Fig. 5.30, where the results are shown in a linear plot. Besides, the value for the interaction range $r_{\rho,\text{GS}}$ is unreasonably small. It takes a value of $r_{\rho,\text{GS}} = 0.01$.

Altogether the FS exponents describe the dynamics on the accessible times, i.e., short and intermediate times after the quench, much better than the GS exponents. As observed in the results for the spinless fermion model oscillations occur due to the finite momentum cutoff of the interaction which are not captured by

Results for the One-Dimensional Model

the bosonization approach.

The results for the two approaches for various values of the interaction strength U can be found in Fig. 5.31. For too small values of $U \lesssim 0.3W$ the differences between the GS and FS power law are too small to distinguish the power laws on the accessible times. But for larger values of the interaction U the FS power law fits well to the numerical data while the GS power law does not fit the slope of the jump towards longer times.

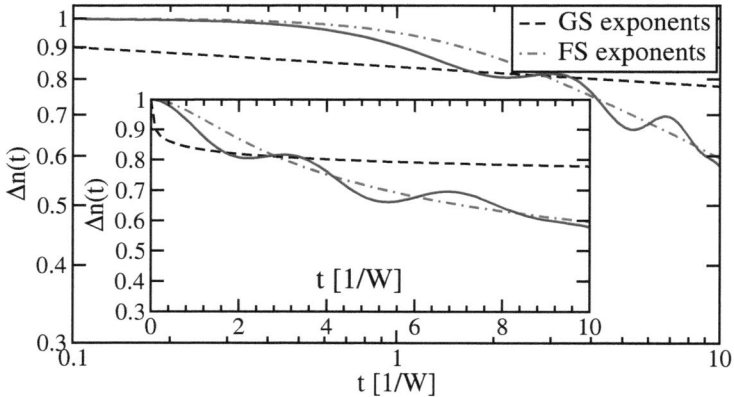

Fig. 5.30.: Jump $\Delta n(t)$ of the quarter-filled Hubbard model with $U = 0.8W$. Dashed lines: Fit to the data with the GS exponents derived by Bethe ansatz with $r_{\rho,\text{GS}} \approx 0.01$. Dashed dotted lines: Fit with the FS exponents from bosonization in the vicinity of the Fermi sea. The ranges are determined to be $r_{\rho,\text{FS}} \approx 1$ and $r_{\sigma,\text{FS}} = 0.6$.

For interaction strengths of the order of the bandwidth $U \gtrsim W$ the agreement between the numerical data and the FS power laws deteriorates. This can be understood by the breakdown of the Tomonaga-Luttinger description. In the Tomonaga-Luttinger description the behavior is described in terms of interaction-free bosons neglecting backscattering processes and the curvature of

5.3 Bosonization

the dispersion. But for large quenches high excitation energies are induced taking the system far from equilibrium.

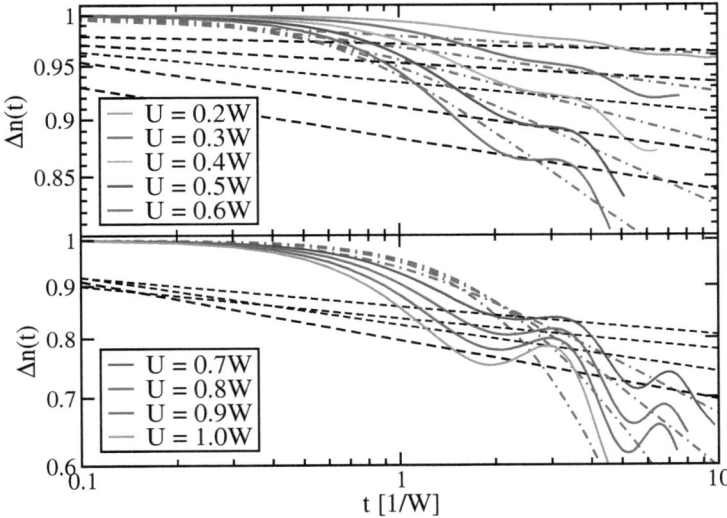

Fig. 5.31.: Jump $\Delta n(t)$ for the quarter-filled Hubbard model for different interaction strengths U. Solid lines: Results of the equation of motion approach. Dashed lines: Fit to the data using the GS exponents. Dashed-dotted lines: Fit by the use of the FS exponents.

In such a situation the effects mentioned above are no longer negligible. Nevertheless on the accessible times these effects are of reduced relevance. This can be seen for instance in the case of backscattering. The leading correction arising from this process is $\propto \int \cos(\sqrt{8}\Phi(x))dx$. On calculating the expectation value with respect to the Fermi sea of this term or one of its derivatives, fluctuations $\langle \Phi(x)^2 \rangle$ occur. These fluctuations diverge so that the cosine term is smeared out to zero and the term vanishes. A more detailed discussion concerning this effect can be found in Ref. [99].

5.3.4.2. Quench energy

For a calculation of the quench energy the ground state energy of the quenched Hubbard model is derived by the use of the Bethe ansatz equations [71]. With the density function $\rho(k)$ the ground state energy of the quenched model can be calculated by

$$E_0 = -\frac{1}{2}N_a \int_{-Q}^{Q} \rho(k)\cos(k)\mathrm{d}k \tag{5.37}$$

with Q determining the integral boundaries $0 < Q \leq \pi$. In the Bethe ansatz solution the integral boundaries determine the particle number n via

$$\int_{-Q}^{Q} \rho(k)\mathrm{d}k = n \tag{5.38}$$

as explained in App. C.

The expectation value of the quenched Hamiltonian with respect to the Fermi sea can be calculated similar to the spinless fermion model. As test for the method the energy of the half-filled model is considered. Corresponding results are shown in Fig. 5.32. For small U the energies increase quadratically with a prefactor of roughly 0.67 in accordance with the second order corrections to the ground state energy [138]. The resulting curve is included in the lower inset of Fig. 5.32. The upper inset shows a linear fit to the large U regime. The curve corresponds to $y = \frac{U}{4} - \frac{1}{\pi}$.

The quench energies for the quarter-filled Hubbard model are shown in Fig. 5.33. For small values of U the curve increases quadratically. However the evaluation of the Bethe ansatz equations in this regime is fairly demanding so that the precision is not the best. For large values of the interaction the results are very accurate.

5.3 Bosonization

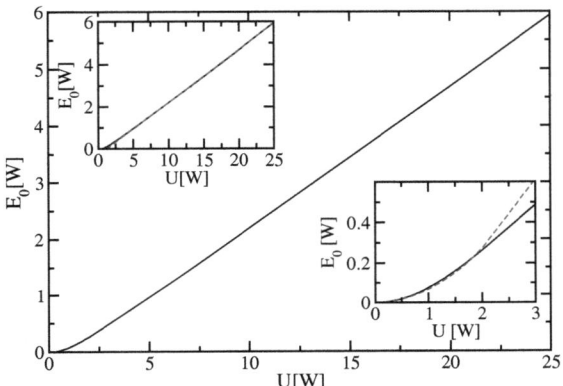

Fig. 5.32.: Quench energy of the half-filled Hubbard model in dependence on the interaction strength U. Black curves: Results derived in a Bethe ansatz calculation according to Eq. 5.37. Red dashed curves: Fits to the data.

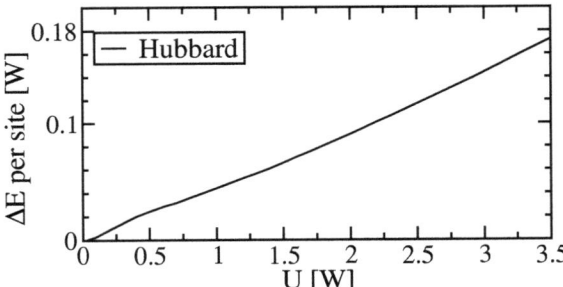

Fig. 5.33.: Quench energy for the quarter-filled Hubbard model. For large U the results of the Bethe ansatz are very accurate whereas the evaluation of the Bethe equations for small U is very difficult, so that slight inaccuracies may occur.

Results for the One-Dimensional Model

From this figure a quench energy of about $0.036W$ can be deduced for a quench to $U = 0.8W$. Comparing the results for the spinless fermion model to the ones for the Hubbard model larger quench energies are more easily accessible in the latter one. Since larger quench energies imply also larger distances to the equilibrium behavior, effects of a larger distance to the equilibrium are more easily accessed in the Hubbard model.

5.3 Bosonization

5.3.5. Relevance of the Results for Longer Times

As illustrated above, the FS exponents describe the behavior of both microscopic models well on the accessible times. But it is still an open question how the systems behave on longer time scales since to date neither analytical techniques nor numerical studies exist which can access the long time behavior of such systems far from equilibrium. On the analytical side a renormalization group theory for non-equilibrium systems is called for.

There is a study of the sine-Gordon model by the use of Keldysh-Green functions [92, 101], but in this work the quench scenario is different from the one used in this thesis. Therefore it is still unclear whether the results can be used to explain the behavior discussed before. In Refs. [92, 101] the bosonic system is quenched at $t = 0$. Afterwards a slow switch-on of the sine term is applied.

For the behavior of the two models on longer times the different cutoffs in the model are essential. These are discussed below. As explained above, there is the $\frac{1}{t}$ cutoff due to the spreading of correlations after the quench with $|\vec{r}| = v_{\max} t$. With the exponentially small energy scale ϵ necessary to detect $K_{\sigma,\text{GS}} \to 1$, the equal-time Green function has to be known at two points with a relative distance $|\vec{r}|$ governed by $\epsilon \approx \frac{v_F}{|\vec{r}|}$. Thus a cutoff $\propto \frac{1}{t}$ is introduced. The accessible times in the equation of motion approach lead to values of about $0.1W$ for the cutoff, depending on the model, the filling factor and the interaction U. From this fairly large cutoff it can be expected that the observed power laws still change gradually for times longer than the ones observed.

Another cutoff is introduced based on the distance of the quenched system to its ground state, as measured by the excitation energy ΔE per site. The energy ΔE per site may be seen as cutoff as the large distance to the equilibrium hinders the system from reaching lowest energies. Thus the flow of the parameters of the effective model is stopped before their fixed point at lowest energies is reached.

Besides the energy $\Delta E/L$ itself, the cutoff scale might also be set by a fictitious temperature T_{quench}, given as the temperature needed to induce the energy $\Delta E/L$ by thermal fluctuations [94, 139]. Such a

Results for the One-Dimensional Model

temperature would account for a relaxation of the model on moderate times. But up to now such a relaxation has not been observed. On the contrary there is growing evidence that integrable models do not thermalize [38, 60, 99, 100]. The steady state of these models is described by a generalized Gibbs ensemble [28].

The above considerations suggest potential crossovers of the dynamics due to the different cutoffs.

i) On short times after the quench the system is described by power laws with FS exponents as derived here.

ii) On intermediate times the behavior is governed by slowly varying exponents which flow towards the GS exponents but never reaching them. Before the GS exponents are reached the distance to the equilibrium $\Delta E/L$ or T_{quench} stops the flow.

iii) On very long times the system ends up in a state where the low energy modes relax towards thermal Gibbs ensembles or towards generalized Gibbs ensembles.

At present no satisfying study of the behavior on long times is possible. Further work is called for to elucidate the precise scenario.

5.4. Periodicity

5.4.1. Strong Quenches

Besides the limit of one dimension the other interesting limit is the limit of infinite dimensions ($d = \infty$). Both limits allow approximation free studies.
Infinite-dimensional models are accessible by dynamical mean field theory (DMFT) [58, 138, 140], which has been reformulated to describe non-equilibrium situations [56, 57]. Eckstein et al., applied non-equilibrium DMFT to the half-filled Hubbard model defined on the infinite-dimensional Bethe lattice [45]. The Bethe lattice has a semielliptical density of states given by $\rho(\omega) = \frac{4}{\pi W} \sqrt{W^2 - 4\omega^2}$.

In this section the results of their approach are compared to the results for the one-dimensional Hubbard model derived within this thesis. Surprisingly, the results show some similarities although the underlying models have totally different inherent properties. In contrast to the infinite dimensional model the one-dimensional model is integrable, i.e., a macroscopic number of conserved quantities exists [129], which influences the dynamics strongly [28]. Furthermore the two models differ regarding scattering processes between their excitations. In the one-dimensional model these processes are controlled by momentum conservation [85–87]. For large dimensions momentum conservation is suppressed at internal vertices [140]. Due to these differences the common lore expects qualitatively different relaxation behavior for the two quenched models.
A comparison of the jump for the two models can be found in Figs. 5.34 and 5.35.
For values of the interaction below the band width $U \lesssim W$ quantitative differences between the two models prevail beyond $t = 2/W$. But for large U surprising similarities occur (see Fig. 5.35).

Results for the One-Dimensional Model

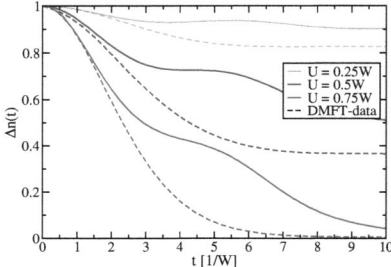

Fig. 5.34.: Jump for various values of the interaction U derived in the equation of motion approach with 11 loops compared to the DMFT-data [59].

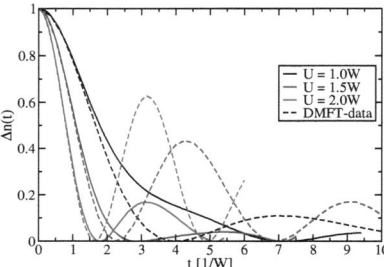

Fig. 5.35.: Comparison of the jump derived in 11 loops (solid lines) to the data derived by DMFT (dashed lines) for larger interactions U.

For quenches to $U = 2.0W$ both systems show coherent oscillations with minima touching zero.

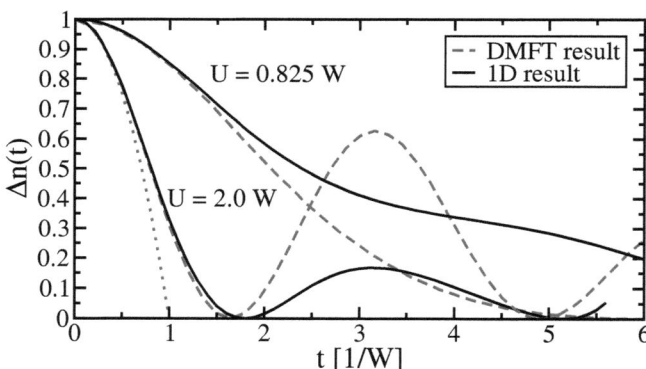

Fig. 5.36.: Solid black lines: Jump $\Delta n(t)$ for the one-dimensional Hubbard model. Dashed lines: Results for the $d = \infty$ Hubbard model on the Bethe lattice, taken from Ref. [45]. Dotted lines: Leading order perturbation theory result $\Delta n(t) = 1 - \frac{U^2}{4}t^2$.

5.4 Periodicity

The minima appear at almost the same instants of time t. These similarities go beyond the leading order perturbation theory in t given by

$$\Delta n(t)_{\text{pert}} = 1 - \frac{U^2}{4}t^2 + O(t^4) \qquad (5.39)$$

derived from Eq. 5.6. The corresponding curve is included in Fig. 5.36 as dotted green line. For large U with $U > W$ and especially for $U = 2.0W$ the positions of the minima show similarities whereas the amplitudes of the oscillations differ significantly.

To sum it up, the periods of the oscillations behave similarly whereas the amplitudes are determined by the dimensionality of the model.

For large U the periods can be read off from the first minima of the curves. In contrast to many other methods the iterated equation of motion approach allows to study arbitrarily large interaction strengths. For strong quenches the equation of motion approach is well controlled, as the number of operators created by commutations with the interaction term H_{int} is restricted by the size of the Hilbert space. In a calculation concerning a fixed number of sites the Hilbert space also stays finite leading to a finite number of operators.

Exemplary results for quenches to large U are shown in Fig. 5.37. It can be seen how the period is decreased on increasing U. Resulting values for the periods in the half-filled and the quarter-filled one-dimensional Hubbard model are shown in Fig. 5.38. Additionally the figure contains the curve $T = \frac{2\pi}{U}$. This value for the period corresponds to the period found in Rabi oscillations as explained below.

Both curves approach the $T = \frac{2\pi}{U}$ curve for large interactions. Thus for large U the physics is governed by local Rabi oscillations. These occur in the two-level system of a singly occupied site and a doubly occupied site containing a particle and a particle-hole pair. This process takes place on one single site and thus it is completely local. Consequently the lattice behaves like a lattice of independent Hubbard atoms in the first approximation which explains the independence of the period of the lattice structure and

Results for the One-Dimensional Model

the dimension of the underlying lattice.

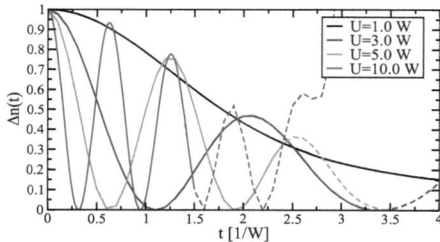
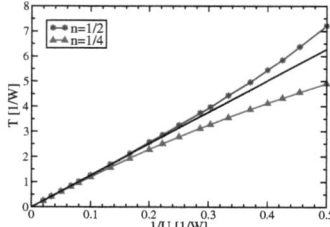

Fig. 5.37.: Exemplary results for the jump $\Delta n(t)$ for strong quenches. The curves show coherent oscillations with a period decreasing with increasing U.

Fig. 5.38.: Period of the oscillations for the half-filled and the quarter-filled Hubbard model in dependence on the inverse interaction strength $\frac{1}{U}$. The black line symbolizes $T = \frac{2\pi}{U}$, the period of local Rabi oscillations.

The appearance of Rabi oscillations can be understood from a local 2-loop calculation, considering the monomials $\hat{c}^\dagger_{0,\uparrow}$, $\hat{c}^\dagger_{-1,\uparrow}$, $\hat{c}^\dagger_{1,\uparrow}$ and the local doubly occupied $\hat{c}^\dagger_{0,\uparrow}\hat{c}^\dagger_{0,\downarrow}\hat{c}_{0,\downarrow}$. With these monomials the time dependent operator $\hat{c}^\dagger_{0,\uparrow}(t)$ is given through

$$\hat{c}^\dagger_{0,\uparrow}(t) = h_0(-1,t) : \hat{c}^\dagger_{-1,\uparrow} : + h_0(0,t) : \hat{c}^\dagger_{0,\uparrow} : + h_0(1,t) : \hat{c}^\dagger_{1,\uparrow} :$$
$$+ h_1(0,0,0,t) : \hat{c}^\dagger_{0,\uparrow}\hat{c}^\dagger_{0,\downarrow}\hat{c}^\dagger_{0,\downarrow} : \quad (5.40)$$

where site 0 represents any site on the lattice due to translational invariance. The corresponding differential equations read

$$\partial_t h_0(0,t) = -Jih_0(-1,t) - Jih_0(1,t) + Un(1-n)ih_1(0,0,0,t), \quad (5.41a)$$
$$\partial_t h_0(-1,t) = -Jih_0(0,t), \quad (5.41b)$$
$$\partial_t h_0(1,t) = -Jih_0(0,t), \quad (5.41c)$$
$$\partial_t h_1(0,0,0,t) = Uih_0(0,t) + (1-2n)Uih_1(0,0,0,t). \quad (5.41d)$$

5.4 Periodicity

At half-filling ($n = 0.5$) the set of differential equations can be solved analytically, resulting in $T = \frac{2\pi}{\sqrt{U^2 + \frac{W^2}{2}}}$. For half-filling the result can be expanded in terms of $\frac{1}{U}$ which yields

$$T = \frac{2\pi}{U} + O\left(\frac{1}{U^3}\right). \tag{5.42}$$

Away from half-filling it can easily be solved numerically. In both cases $T = \frac{2\pi}{U}$ represents the leading order result in $\frac{1}{U}$.

Additional commutations with the interaction term will not create new monomials acting on the considered site 0. Consequently the 2-loop calculation captures the leading order in t independent of U. It becomes exact in the limit $U \to \infty$ as illustrated in Fig. 5.38. For increasing U the curves approach the $T = \frac{2\pi}{U}$ curve representing local Rabi oscillations.

The results of the 2-loop calculation presented here are only valid in leading order. Therefore the results presented in Fig. 5.38 deviate from the results expected from this 2-loop calculation in higher orders.

A calculation covering the subleading order $\frac{W}{U^2}$ has to include all operators acting on two sites. These would be included in a 7-loop calculation with 2210 monomials. Thus an analytic analysis of the subleading term is not possible.

In conclusion, the surprising similarities to the DMFT data illustrate that the physics for large quenches is governed by local processes. This can be understood analytically by the iterated equation of motion approach developed in this thesis. On the level of such a calculation, the surrounding lattice acts as a damping bath so that details of the lattice are of minor importance.

5.4.2. Dynamical Transition

For the half-filled, $d = \infty$ Hubbard model the DMFT results revealed two regimes depending on the interaction strength U [45]. In the two regimes the jump behaves qualitatively different. For weak quenches the jump gradually decreases to a prethermalization plateau. Exposed to strong quenches the jump oscillates strongly with minima reaching $\Delta n = 0$ (see Fig. 5.35). The two regimes are separated by a **dynamical** transition.

A dynamical transition is characterized by the observation that already a small change in the corresponding parameter leads to a dramatic change in the dynamics as observed in various models like the Bose-Hubbard model [141] and the transverse field Ising model [142].

This transition has also been observed in a variational Gutzwiller approach for a quenched Hubbard model [143] as well as for a model with a linear ramp of the interaction [144]. In the Gutzwiller approach the dynamics of the Hubbard model is mapped to a classical mechanics problem which can then be solved analytically. The faith of an analytical solution is paid off by neglecting quantum fluctuations [145]. Thus true relaxation effects are not captured by this approach. For both regimes oscillations $\propto \cos\left(\frac{2\pi}{T}t\right)$ are observed, with the period T

$$T = \begin{cases} \frac{4\sqrt{2}}{U_c}K(2U/U_c) & \text{for } U < U_c \\ \frac{4}{U_c}K(U_c/2U) & \text{for } U > U_c, \end{cases} \quad (5.43)$$

for the half-filled model. In the above formula $K(x)$ denotes the complete elliptic integral of the first kind [146] and U_C labels the interaction for which the Mott transition is expected in the Gutzwiller approach [147]. It can be calculated by the kinetic energy per particle E_{kin}/N via $U_C = -8E_{kin}/N$. In the Gutzwiller method the two regimes are separated by a singularity in the oscillation period. Upon doping the transition turns into a crossover. As the Gutzwiller approach becomes exact in the limit of infinite dimensions ($d \to \infty$) [148, 149], similarities between the Gutzwiller results and the DMFT results are

5.4 Periodicity

not surprising.

But the models used in these approaches show qualitatively different properties from the one-dimensional Hubbard model as explained in Sect. 5.4.1. Nevertheless, results of the iterated equation of motion approach for the one-dimensional Hubbard model also exhibit two regimes, see Fig. 5.34 and Fig. 5.35. For weak quenches, the jump shows oscillations and is only gradually decreasing. On increasing U the decrease becomes stronger until some kind of shoulder occurs as can be seen, for instance, in the curve for $U = 0.75W$ of Fig. 5.34. Increasing U further leads to a decay with nearly no oscillations for $U = 1.25W$. For even larger U the pronounced oscillations discussed in Sect. 5.4.1 dominate. Clearly the period of the oscillations changes with U. This observation can be used to determine the dynamical transition.

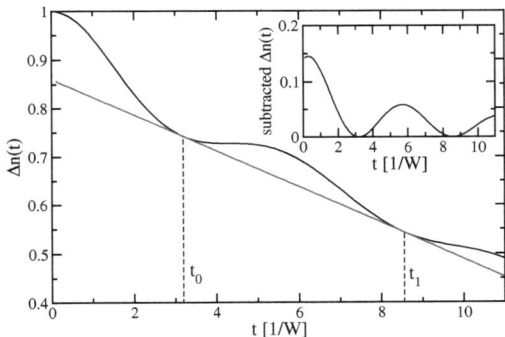

Fig. 5.39.: Jump for the half-filled model with an interaction $U = 0.5W$. Red line: Corresponding fit used to capture the slope of the jump with the positions of the first minima denoted by t_0 and t_1. Inset: Curve for the jump substracted by the fitted slope. The period of the oscillations can clearly be seen.

The iterated equation of motion approach incorporates true relaxation effects. Thus the periods cannot simply be read off as it is the case in the relaxation-free Gutzwiller approach, where the am-

Results for the One-Dimensional Model

plitudes of the oscillations are preserved. For not too large U the oscillations are superseded with a decay, which has to be taken into account. To capture the relaxation, a straight line is fitted to the data as a double tangent to two points close to the first minima. Such a tangent for a quench to $U = 0.5W$ is displayed in Fig. 5.39. The curve determined by subtracting the fitted line from the original jump $\Delta n(t)$ is shown in the inset of this figure.

Then the period is given by $T = 2t_0$, where t_0 denotes the first of the two points in Fig. 5.39. This fit procedure is useful especially for small U, where only small oscillations appear on top of the decay. For values of U where only one minimum lies within the range of convergence (for instance $U = 1.5W$) the first minimum is used to determine t_0.

The crossover from using a fit to reading the minimum directly is set by the behavior of the second derivative of the jump. This is shown in Fig. 5.41 for some U in this range.

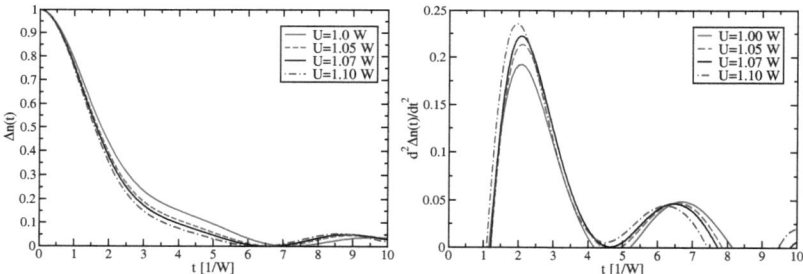

Fig. 5.40.: Jump for the half-filled model for various values of U. In this range the jump exhibits a kind of shoulder which vanishes for larger interaction.

Fig. 5.41.: Second derivative of the jump $\Delta n(t)$ with time in the interaction range where the crossover from a shoulder to a pure decrease takes place.

It can be seen how the shoulder visible in the curve for $U = 1.0W$ vanishes gradually on increasing U, leaving behind a pure decrease of the jump. If two turning points exist, i.e., two zeros in the second

5.4 Periodicity

derivative ($U < 1.07W$), the fit procedure is used. The resulting periods are given in Fig. 5.42 as function of the compactified interaction $\frac{U}{U_C/2+U}$. As explained, U_C is given by $U_C = -8E_{kin}/N$ with the kinetic energy E_{kin}/N. The kinetic energy is calculated by the expectation value of the initial Hamiltonian

$$E_{kin}/N = -\frac{J}{N}\sum_{\sigma}\sum_{<i,j>}\langle\hat{c}^\dagger_{i,\sigma}\hat{c}_{j,\sigma}\rangle \qquad (5.44)$$

where the sum $<i,j>$ runs over nearest neigbors. Due to the spin and mirror symmetry the expectation values are reduced to $\langle\hat{c}^\dagger_{i,\uparrow}\hat{c}_{i+1,\uparrow}\rangle = -\frac{1}{\pi}$ which yields $E_{kin}/N = J\cdot 2\cdot 2\left(-\frac{1}{\pi}\right) = -\frac{1}{\pi}W$ as the hopping element contributes $J = \frac{1}{4}W$. Consequently the value for the interaction U_C reads $U_C = \frac{8W}{\pi}$.

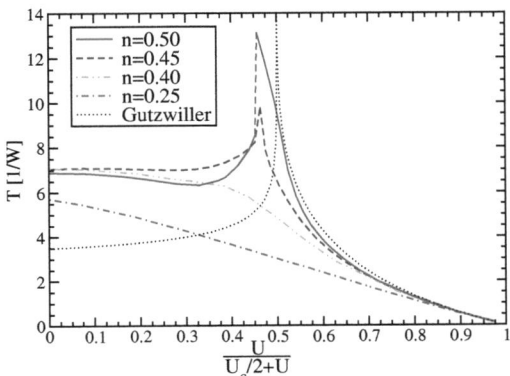

Fig. 5.42.: Period of the oscillations dependent on U with $U_C = 8W/\pi$ in one dimension. The curves are shown for various fillings. Additionally the result of the semiclassical Gutzwiller approach [143] is included as dotted line.

The resulting periods are plotted in Fig. 5.42. At half-filling and for small values of the interaction ($U \lesssim U_C/2$) the period shows hardly

Results for the One-Dimensional Model

any dependence on U. This is in qualitative agreement with the Gutzwiller results, where the curve stays flat in this regime. But the values of T differ from the ones obtained by the Gutzwiller approach by a factor of two. The value reached in the iterated equation of motion approach for $U \to 0$ depends on the filling factor n. A rough suggestion for the period T at $U = 0$ is $T = \frac{2\pi}{W}$. For values of $\frac{U}{U_C/2+U} \leq 0.35W$ the period is first decreased on increasing U. For larger values of the interaction the period is increased in the regime of weak quenches.

For strong quenches the period decreases on increasing U until the $T = \frac{2\pi}{U}$ curve is reached, due to the emergence of local Rabi oscillations. The two regimes are separated by an anomaly catching the eye in Fig. 5.42 at about $U = U_C/2$. With the equation of motion approach times larger than $t \approx 15/W$ cannot be reached for this model. Thus it cannot be determined whether the anomaly is a sharp peak as indicated by the dashed line or a singularity. However, the data shows an anomalous behavior with periods rising to about twice their initial value. In the Gutzwiller approach a singularity is found. The periods derived by this approach are depicted as dotted curves in Fig. 5.42. The curves agree well for the large U limit where generic Rabi oscillations are found independent of the dimension of the underlying lattice. In contrast to the results of this thesis the Gutzwiller results increase for $U \leq 0.35W$ on increasing U. Besides the position of the anomaly is slightly shifted from $U = U_C/2$ in the Gutzwiller data to $U \approx 0.43 U_C$ in the equation of motion approach.

Comparing Fig. 5.34 and Fig. 5.35 it can be noticed that for all curves to the right of the anomaly, the minima of the curves reach zero, whereas the curves to the left display oscillations without zeros. This represents an easy to detect qualitative fingerprint which distinguishes both regimes. Upon doping this difference is lost. As shown in Sect. 5.2 the oscillations do not exhibit zeros even for very large values of U ($U = 100W$) for doped systems. Even for smallest values of the doping the transition ceases to exist.

5.4 Periodicity

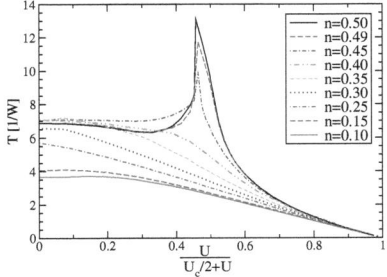

Fig. 5.43.: Period of oscillations for various fillings in dependence on U with n denoting the filling factor of one spin species.

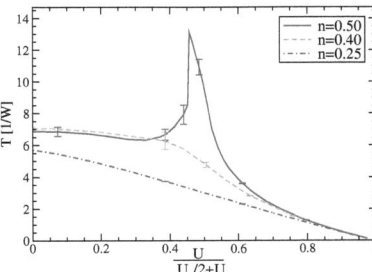

Fig. 5.44.: Periods for various fillings. The error bars depict exemplary results for the estimated errors in the determination of the periods.

Periods determined for various filling factors n are depicted in Fig. 5.43. Note that n is the filling factor of one spin species. Upon doping the anomaly is gradually washed out. About 20% of doping is needed to make it vanish completely.

In the determination of the periods inaccuracies occur due to the finite number of loops performed, i.e., the finite convergence range, and due to the fitting procedure. The effect of the first error source is estimated by calculating the period for different number of loops. The difference of these values is then used as an estimate for the systematic error. The errors due to the linear fit used to determine the period in the weak quench regime are calculated by determining the period additionally on a different route. For the values of U where the fitting procedure is used, the position of the first minimum $\tilde{t}_0$ is directly read off to calculate $\tilde{T} = 2\tilde{t}_0$. Then the error is bounded by the absolute difference $|T - \tilde{T}|$. Finally the maximum of both estimates is used as the error of the corresponding period. In the cases where T is deduced from a double tangent fit the difference $|T - T^*|$ with $T^* = t_1 - t_0$ can be used as additional estimate for the error. This estimate yields values of the period which are shorter by up to $1/W$ for small U. For some exemplary results the value of the error derived as maximum of the different estimates is included

Results for the One-Dimensional Model

in Fig. 5.44. Large interactions lead to very small errors so that the error bars can hardly be seen, see for instance the half-filled curve for $\frac{U}{U_C/2+U} = 0.8W$ or the $n = 0.25$ curve. Of course, the errors are largest in the vicinity of the transition. But the slope of the peak and its position is hardly affected so that the conclusions stay valid.

In conclusion, the one-dimensional Hubbard model also shows a dynamical transition [150], so that it can be stated that the transition observed in the Gutzwiller approach is not due to the semiclassical approximations. Since two Hubbard models with very distinct dimensionality show this transition, it can be declared a generic feature of quenched Hubbard models. Further evidence will be provided in two dimensions (see Sect. 6.1.5.2).

6. Two-dimensional Model

Up to now studies of systems far from equilibrium are mostly focussed on one-dimensional models, infinite-dimensional models or small finite systems, because for these many theoretical techniques are available. One-dimensional systems are, for example, accessible by quantum field theories [91] or time dependent density matrix renormalization group approaches [26, 50, 51]. Infinite-dimensional models are studied by the use of dynamical mean-field theories [45, 56, 57] and Gutzwiller approaches [143]. For small finite systems exact diagonalization is possible [23, 151].

In contrast to this only a few studies of two-dimensional systems out of equilibrium exist. Continuous time quantum Monte Carlo (QMC) was used to study the two-dimensional Hubbard model on a 20×20 lattice. In this QMC-study the quench protocol starts with an interacting model, quenching the system to the interaction-free model [63]. In contrast to the quench protocol studied in this thesis the system starts in a rather complicated state and is quenched to a simple Hamiltonian. Thus the resulting dynamics is rather well-understood by perturbative approaches as the physics is governed by the correlations of the initial state in this quench protocol.

Other studies of two-dimensional models are based on recursively constructed Hilbert spaces. These studies concern single charge carriers in a Mott insulator exposed to a strong electric field [152] and a bound pair of two charge carriers [153].

In this chapter the Hubbard model is considered on a two-dimensional square lattice. This model is not integrable and other peculiarities of the one-dimensional model are absent. There is, for example, no dominant momentum conservation, so that this model can be viewed as a generic model with finite dimension ($\neq 1$). In this model true relaxation is expected to occur, which makes a study of its dynamics especially attractive.

Two-dimensional Model

6.1. Half-Filled Two-Dimensional Model

In the square lattice the coordination number is $z = 4$ and consequently the band width is $W = 8J$ with the hopping parameter J. The corresponding dispersion reads

$$\epsilon_k = -2J(\cos(k_x) + \cos(k_y)) \tag{6.1a}$$
$$= -4J\cos\left(\frac{k_x+k_y}{2}\right)\cos\left(\frac{k_x-k_y}{2}\right). \tag{6.1b}$$

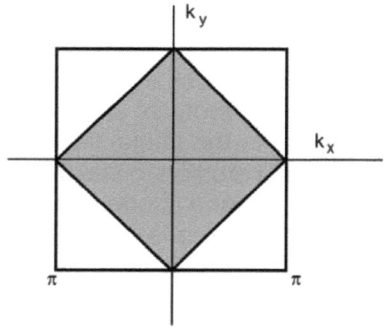

Fig. 6.1.: Fermi surface of the half-filled Hubbard model defined on a two-dimensional square lattice.

At half-filling the Fermi energy vanishes $\epsilon_F = 0$. Thus the Fermi surface is determined by $\cos(k_x) = -\cos(k_y)$. Consequently the momenta have to fulfill

$$|k_x| + |k_y| \leq \pi \tag{6.2}$$

resulting in a box shaped Fermi surface as shown in Fig. 6.1.

The density of states for the two-dimensional model is given through [154, 155]

$$\rho(\epsilon) = \frac{1}{2\pi^2|J|}\mathcal{K}\left(\sqrt{1-\frac{\epsilon^2}{16J^2}}\right) \tag{6.3}$$

6.1 Half-Filled Two-Dimensional Model

with the complete elliptic integral of the first kind [146] defined as

$$\mathcal{K}(z) = \int_0^1 \frac{1}{\sqrt{1-w^2}} \frac{1}{\sqrt{1-z^2 w^2}} dw \quad (6.4)$$

with $\mathcal{K}(z) = \mathcal{K}(-z)$. For vanishing z the integral yields $\mathcal{K}(0) = \frac{\pi}{2}$ and in the limit $\lim_{z\to\infty} \mathcal{K}(z) = 0$.
Most notably are the step functions for $\epsilon = \pm 0.5\pi$ and the logarithmic divergence, which appears for $\epsilon = 0$.

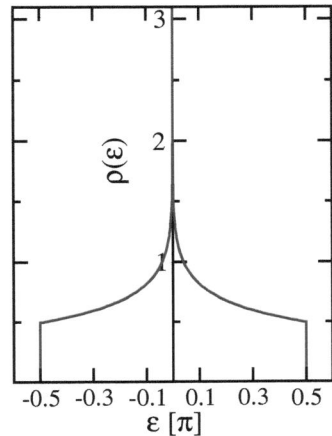

Fig. 6.2.: Density of states for the half-filled two-dimensional Hubbard model.

As explined in Sect. 3.2.1 the correlations are needed in order to apply normal ordering. To simplify the calculation of the contractions the position of an operator on the two-dimensional lattice is labelled by two indices $\vec{r} = (l, j)^T$, so that $\langle c^\dagger_{\vec{r},\uparrow} c_{\vec{0},\uparrow} \rangle = \langle c^\dagger_{lj,\uparrow} c_{00,\uparrow} \rangle$. Just like in the one-dimensional case the contractions can be calculated via

$$\langle c^\dagger_{lj,\uparrow} c_{00,\uparrow} \rangle = \frac{1}{(2\pi)^2} \int_{-\infty}^{\infty}\int_{-\infty}^{\infty} e^{i(k_x l + k_y j)} \underbrace{\delta(\vec{k}-\vec{k'})\Theta(\epsilon_F - \epsilon_k)}_{\langle c^\dagger_{\vec{k}} c_{\vec{k'}} \rangle} d^2k d^2k' \quad (6.5a)$$

$$= \frac{1}{(2\pi)^2} \int \cos(k_x l + k_y j)\Theta(\epsilon_F - \epsilon_k) d^2k \quad (6.5b)$$

with the dispersion ϵ_k.

Two-dimensional Model

To simplify the calculations relative coordinates are introduced

$$s = \frac{k_x + k_y}{2} \quad \text{und} \quad p = \frac{k_x - k_y}{2} \tag{6.6a}$$

$$\text{which lead to} \quad k_x = s+p \quad k_y = s-p \tag{6.6b}$$

$$dk_x dk_y = 2 ds dp. \tag{6.6c}$$

With the new coordinates the integral is converted according to

$$\langle c^\dagger_{lj,\uparrow} c_{00,\uparrow} \rangle = \frac{1}{(2\pi)^2} \int \int_{-\frac{\pi}{2}}^{\frac{\pi}{2}} 2\cos((s+p)l + (s-p)j) ds dp \tag{6.7a}$$

$$= \frac{1}{(2\pi)^2} \int \int_{-\frac{\pi}{2}}^{\frac{\pi}{2}} 2\cos((s+p)l)\cos((s-p)j) ds dp \tag{6.7b}$$

$$= \frac{2}{\pi^2} \frac{\sin\left(\frac{\pi}{2}(l+j)\right) \sin\left(\frac{\pi}{2}(l-j)\right)}{(l+j)(l-j)}. \tag{6.7c}$$

Having determined the correlations the iterated equation of motion approach can be applied to determine the momentum distribution.

6.1 Half-Filled Two-Dimensional Model

6.1.1. Convergence

In the two-dimensional model more hopping processes are possible so that much more monomials are created within one loop than in the one-dimensional model. The number of monomials increases exponentially with a factor of about 6.6 for each loop. This factor has to be compared to a factor of about 3 which is observed for studies of the one-dimensional model. Consequently in an iterated equation of motion approach for the two-dimensional model less commutations are feasible than for the one-dimensional model. In the two-dimensional case up to nine loops are possible, whereas 11 loops are performed for the one-dimensional model. As check for the convergence, results for calculations with differing numbers of commutations are performed. The absolute difference of these results for a quench to $U = 0.5W$ is depicted in Fig. 6.3 in a double logarithmic plot, where the result of the 9-loop calculation is used as a reference.

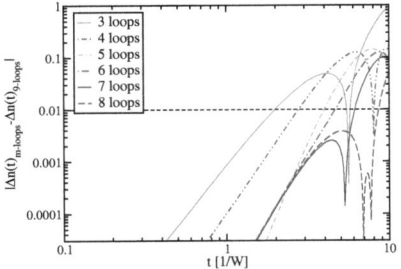 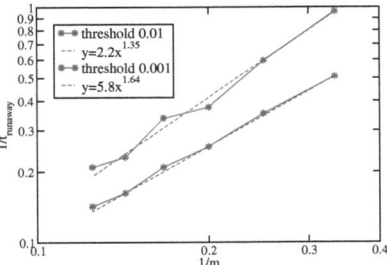

Fig. 6.3.: Absolute difference in the results for the jump with $U = 0.5W$ derived in calculations with different numbers of loops m in a double logarithmic plot. The result of a 9-loop calculation is used as a reference.

Fig. 6.4.: Inverse runaway time for a quench to $U = 0.5W$ in dependence on $1/m$ with m denoting the number of loops performed (see also Sect. 3.5). The thresholds are 0.01 and 0.001 which yield exponents of about 1.35 and 1.64 respectively.

The inverse runaway time as defined in Sect. 3.5 is shown in Fig. 6.4 in a double logarithmic plot with the inverse number of loops

$1/m$ given on the x-axis. As explained, the choice of the threshold is arbitrary. However, thresholds of 0.01 and 0.001 are chosen to illustrate the increase in the range of convergence with increasing number of loops m. Both curves show that the inverse runaway time vanishes for an infinite number of loops, indicating that the results become exact for all times in this limit. Fits to the data reveal an exponent of 1.35 for a threshold of 0.01 and an exponent of about 1.64 for a threshold of 0.001. Thus the increase is superlinear like in the one-dimensional model, although the exponents are slightly smaller.

Even though less loops are performed in the studies of the two-dimensional model, the range of convergence is comparable to the ones observed in the results for the one-dimensional model. For a quench to $U = 0.3W$ the results are converged for times $t \lesssim 8/W$ which is comparable to the times reached in the one-dimensional case.

6.1.2. Results for the Half-Filled Two-Dimensional Model

The results presented in the following are derived at the edge of the Fermi surface at $(0, \pi)$. A comparison of the behavior of the jumps calculated at different points on the Fermi surface is provided in Sect. 6.1.6. On the accessible times there is no sizeable dependence of the jump on the position on the Fermi surface. Even the curves calculated at the edges of the Fermi surface $(\pm\pi, 0)$ and $(0, \pm\pi)$ behave similarly to those calculated in the middle at $(\pm\frac{\pi}{2}, \pm\frac{\pi}{2})$ on the accessible time scales. Thus the results presented in the following for $(0, \pi)$ are representative for jumps all over the Fermi surface.

The corresponding curves for various values of U derived in the iterated equation of motion approach [156] are depicted in Fig. 6.5. The two-dimensional model also shows two distinct regimes very similar to the ones in the one-dimensional model. On the square lattice strong quenches $U \gtrsim 0.7W$ lead to oscillations with zeros as discussed in Sect. 5.4.1. For weak quenches a decay of

6.1 Half-Filled Two-Dimensional Model

the jump can be observed but in contrast to the chain-model only very weak oscillations occur. This observation can be explained by the band edges. In one dimension the oscillations are much stronger due to the pronounced van-Hove singularities [93] (see Sect. 4.3). The much stronger oscillations can be seen in Fig. 6.6, where additionally to the curves for the two-dimensional case the results for the one-dimensional model are included as well. On the other hand, the DMFT results show nearly no oscillations (see App. D), which agrees with the assumption that the oscillations are caused by the band edges, as the Bethe lattice has a semi-elliptical density of states and a dispersion with square root singularities [45] which are even weaker than the singularities in the two-dimensional case. The oscillations due to the band edges are already present in the U^2-calculations in Sect. 4.3.

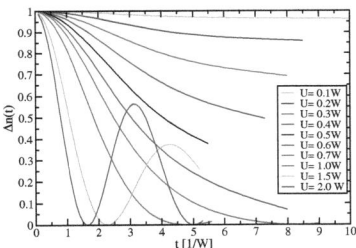

Fig. 6.5.: Jump for the two-dimensional Hubbard model derived in 9-loop calculations. Note the increased range of convergence for intermediate $U \approx 0.7W$.

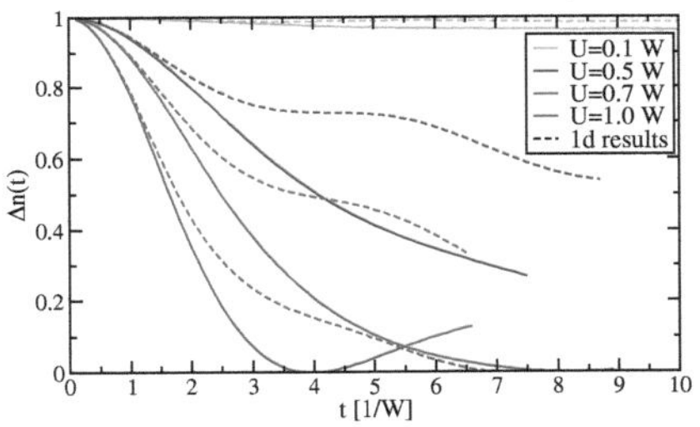

Fig. 6.6.: Solid lines: Results for the jump $\Delta n(t)$ in the two-dimensional Hubbard model. Dashed lines: Corresponding results for the one-dimensional Hubbard model.

Comparing the curves for the one-dimensional model to the ones derived for the two-dimensional model, the curves for the one-dimensional and the two-dimensional model start for small times with the same quadratic behavior.
But for larger times the two-dimensional model shows a much faster relaxation than the one-dimensional model. This difference

Two-dimensional Model

can be explained by the differences in the two models. The two-dimensional model inherits effective scattering mechanisms, absent in one dimension.

As can be seen the range of convergence is drastically enhanced for intermediate interaction strengths ($U \approx 0.7W$). In general the range of convergence decreases on increasing U (see Sect. 3.2.2). In the two-dimensional model similar behavior can be found except for the intermediate U regime, where a fast decay of the jump is observed.

For values of U of about $0.6 - 0.7W$ the data shows an exponential decrease e^{-at} with a relaxation rate a. In the following the relaxation of the jump and the corresponding relaxation rate a is addressed for strong and for weak quenches separately.

6.1.3. Strong Quenches

For large interactions U the curves for the two-dimensional model reveal the same coherent oscillations as the curves for the one-dimensional model, supporting the view that their appearance is a generic feature of strongly quenched Hubbard models as discussed in Sect. 5.4.1. Examples for such oscillations can be found in Fig. 6.5, e.g., for $U = 2.0W$. The oscillations show a decrease in time of their amplitudes. A corresponding relaxation rate is determined in the following by the use of fit functions. To account for the strong oscillations a cosine with frequency ω is used in the fit function. Additionally the fit function includes an exponential function

$$\Delta n_{\text{strong}} = \cos(\omega t)^2 \exp(-\sqrt{(at)^2 + b^2} + b) \tag{6.8}$$

with relaxation rate a. The parameter b is introduced to be able to reproduce the second order in Ut result appropriately. Without this parameter the fit function would exhibit cusps at small t.

6.1 Half-Filled Two-Dimensional Model

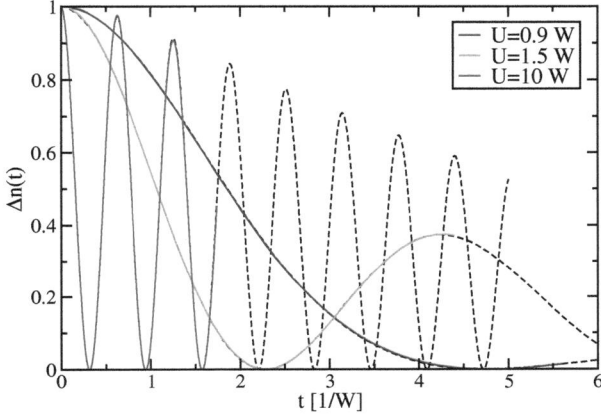

Fig. 6.7.: Jump $\Delta n(t)$ in the strong quench regime in dependence on the time t. Dashed lines: Fits to the data according to Eq. 6.8.

In this way the fit function combines the behavior for small t with the exponential decay expected for larger times. Exemplary results for the fit are depicted in Fig. 6.7. The parameters a and $T = \frac{2\pi}{\omega}$ used in these fits as well as the calculated values for b are given in Fig. 6.8. These are depicted in dependence on the compactified interaction $\frac{U}{U_C/2+U}$, where U_C is determined as described in Sect. 5.4.2 by the kinetic energy

$$\frac{E_{\text{kin}}}{N} = -2J\left(\langle \hat{c}_{i,j}^\dagger \hat{c}_{i,j+1}\rangle + \langle \hat{c}_{i+1,j}^\dagger \hat{c}_{i,j}\rangle + \langle \hat{c}_{i,j+1}^\dagger \hat{c}_{i,j}\rangle + \langle \hat{c}_{i,j}^\dagger \hat{c}_{i+1,j}\rangle\right) \quad (6.9)$$

with the factor of 2 accounting for the spin. Due to the symmetries the expectation values coincide yielding $\langle \hat{c}_{1,0}^\dagger \hat{c}_{0,0}\rangle = \frac{2}{\pi^2}$. Using the identity $J = \frac{1}{8}W$ with the band width W the kinetic energy reads

Two-dimensional Model

$$\frac{E_{kin}}{N} = -\frac{2}{\pi^2}W \quad (6.10a)$$

$$\Rightarrow U_C = -8E_{kin}/N = \frac{16W}{\pi^2}. \quad (6.10b)$$

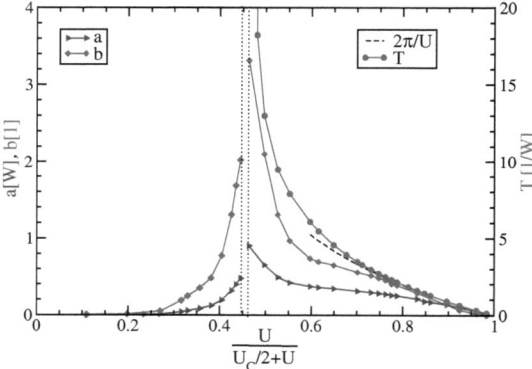

Fig. 6.8.: Fitting parameters a,b and $T = \frac{2\pi}{\omega}$ as function of the compactified interaction $\frac{U}{U_C/2+U}$ with $U_C = \frac{16W}{\pi^2}$. The scale for a and b is shown on the left whereas the scale for T is given on the right. In the range between the two dashed vertical lines the results do not allow for a unique assignment to one of the two regimes. For strong quenches the period approaches the value of Rabi oscillations $\frac{2\pi}{U}$ given as dashed line.

It can be seen that the relaxation rate decreases on increasing U in the strong quench regime. The values shown in this regime start with a value of $U = 0.7W$.

6.1 Half-Filled Two-Dimensional Model

6.1.4. Weak Quenches

For an infinite-dimensional model Moeckel and Kehrein proposed a strict U^2-calculation [46]. The two-dimensional model is more generic than the one-dimensional model and especially it is not integrable, thus an analogue U^2-calculation for this model is of special interest. Results of such a calculation are used as comparison to the 9-loop results. Besides, the results are used in the following for an analysis of the jump in weak quenches. The U^2-calculations can be reproduced by truncating the differential equations in the equation of motion approach.

6.1.5. Strict U^2–Calculations

In contrast to the self-similar U^2-calculation presented in Sect. 4.3 the calculation used in this section is a **strict U^2-calculation** in the sense that there are no higher order terms appearing in the differential equations. In contrast to this the self-similar U^2-calculation presented in Sect. 4.3 is a full calculation in the subspace of monomials describing one- and three-particle terms. Thus also higher orders in U are created when solving the differential equations. In the approach used in this section this is not possible. The results obtained here are strict U^2-results.

To get results comparable to the ones obtained by Moeckel and Kehrein [46], strict results in U^2 are needed. It is not sufficient to truncate the results according to the number of commutations with the interaction as explained in App. B.1, because such a calculation would implicitly contain higher order processes in U (see also Sect. 4.3). The difference between the approaches is explained in the following by the appearance of different terms in the differential equations.

At the beginning only one-particle terms $h(1P)$ are present, which appear in $\mathcal{O}(1)$. For simplicity the corresponding prefactors are regrouped in a vector of coefficients labelled $h(1P)$. By commutation with the interaction new terms with one particle and a particle-hole pair are created and grouped together in the vector $H(2P1H)$. Since these terms are created by commuting with the interaction term $\hat{H}_{\text{int}}$

and obey $H(2P1H) = 0$ for $t = 0$, they are strictly proportional to $\mathcal{U}$. Another commutation with the interaction leads to a feedback effect of these three-particle terms to the one-particle terms. This feedback effect has to be included for results $h(1P)$ correct up to $\mathcal{U}^2$. But additionally this commutation creates three-particle terms similar to the terms contained in $H(2P1H)$. Although these new three-particle terms are similar to the ones in $H(2P1H)$ they must not be considered in a strict $\mathcal{U}^2$-calculation. As $H(2P1H)$ is $\propto \mathcal{U}$, these additional terms lead to $\mathcal{U}^2$-terms for $H(2P1H)$ and thus higher order corrections to $h(1P)$ and the jump. In a strict $\mathcal{U}^2$-calculation these terms have to be neglected, whereas they would be included in a self-similar calculation with a truncation according to the number of commutations with H_{int}. Thus the two approaches differ in the way these higher order terms are treated. To achieve strict $\mathcal{U}^2$-results the influence of the kinetics on $h(1P)$ and of the feedback from the three-particle terms are captured by separate vectors $h(1P)$ and $d(1P)$. The vector $h(1P)$ contains the influence of the kinetics

$$\partial_t h(1P) = \mathcal{T}[h(1P)] \qquad (6.11)$$

with $\mathcal{T}$ denoting the commutation with the kinetic part. The differential equation of the three particle terms reads

$$\partial_t H(2P1H) = \mathcal{T}[H(2P1H)] + \mathcal{U}[h(1p)] \qquad (6.12)$$

containing the kinetics $\mathcal{T}$ and additionally the commutation with the interaction term $\mathcal{U}$. These contributions are $\propto \mathcal{U}$. Commuting the three-particle terms with the interaction leads to corrections to the one-particle terms $\mathcal{U}[H(2P1H)]|_{\text{1-particle}}$, which are captured in the vector $d(1P)$. The corresponding differential equation contains the well-known kinetics

$$\partial_t d(1P) = \mathcal{T}[d(1P)] + \mathcal{U}[H(2P1H)]|_{\text{1-particle}} \qquad (6.13)$$

with the initial conditions $d(1P)|_{t=0} = 0$ leading to $d(1P) \propto \mathcal{U}^2$.

The second order corrections to the jump are given by the product $h(1P)d(1P)$ bilinear in the prefactors. Performed like this the strict $\mathcal{U}^2$-

6.1 Half-Filled Two-Dimensional Model

calculation corresponds to the calculation by Moeckel and Kehrein for the infinite-dimensional model.

Results for the U^2-calculation for two exemplary values of U and different numbers of loops are shown in Fig. 6.9. It can be seen how the range of convergence is increased on increasing loop number. As the amount of terms created in this calculation is limited by the constraints applied to a strict U^2-calculation, far more loops are possible than in a full calculation. The results of the 13-loop calculation correct up to second order U^2 are converged for times up to $t \approx 16/W$.

6.1.5.1. Results

Following Moeckel and Kehrein the jump in a U^2-calculation can be written as [46]

$$\Delta n_{k_F,\text{2nd}} = 1 - U^2 f_{k_F}(t) + O(U^4) \quad \text{with} \tag{6.14a}$$

$$f_{k_F} = \frac{4}{N^2} \sum_{pp'q} \delta^{p'+q}_{p+k_F} \frac{\sin^2(\Delta \epsilon t/2)}{\Delta \epsilon^2} (n_p \bar{n}_{p'} \bar{n}_q + \bar{n}_p n_{p'} n_q) \tag{6.14b}$$

with N denoting the number of sites. The energy difference reads $\Delta \epsilon = \epsilon_{k_F} + \epsilon_p - \epsilon_{p'} - \epsilon_q$ and $n_p = 1$ for p within the Fermi surface. Its complement is denoted by $\bar{n}_p = 1 - n_p$. The function f_{k_F} can also be evaluated by integrating the above formula.

In the strict U^2-calculation presented here f_{k_F} is given by the Fourier transforms of the prefactors $d(1P)$ and $h(1P)$

$$f_{k_F} = d_k^*(1P)h_k(1P) + h_k^*(1P)d_k(1P). \tag{6.15}$$

The result for f_{k_F} as derived in a calculation with 13 loops is shown in Fig. 6.10. Surprisingly it exhibits a logarithmic divergence.

Two-dimensional Model

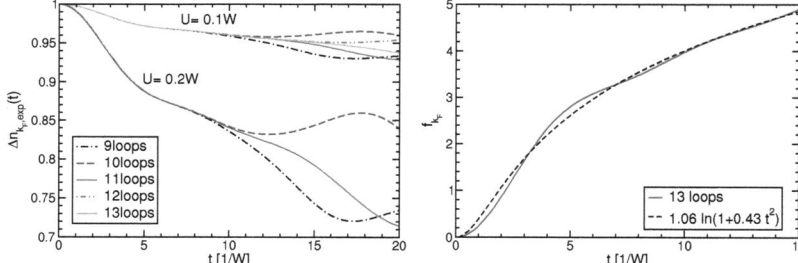

Fig. 6.9.: Jump $\Delta n(t)$ as derived in a strict U^2-calculation for two values of the interaction strength U and various loop numbers.

Fig. 6.10.: Dependence of the function f_{k_F} on time t. Additionally a logarithmic fit to the data is shown indicating a logarithmic divergence.

Additionally this figure contains the curve $1.06\ln(1 + 0.43t^2)$. Apart from oscillations superimposing the increase the behavior of f_{k_F} is very well described by the logarithmic increase. As a consequence, the jump $\Delta n(t)$ does not show a prethermalization plateau in U^2 [46, 48] but decreases without bound becoming eventually negative. This is attributed to the flat shape of the Fermi surface and the shape of the dispersion which is given by flat planes giving rise to nesting effects. Thus the momenta parallel to these planes do not matter and the momenta perpendicular to them behave similar to momenta in one dimension. Consequently the resulting f_{k_F} displays logarithmic divergencies.

In Fig. 6.11 the results obtained by the iterated equation of motion approach are compared to the results of the second order in U approach derived in this section. The curve representing $\Delta n_{k_F, \text{2nd}}(t)$ for a quench to $U = 0.5W$ is included in Fig. 6.11 as dotted line. For small times the curve starts with the desired slope and then takes values smaller than the curve for the full 9-loop calculation. For large values of $t \approx 10/W$ the jump becomes negative. The appearance of negative values holds true even for small values of U if the time is chosen large enough. To circumvent the unphysical negative values

6.1 Half-Filled Two-Dimensional Model

due to the logarithmic divergence of f_{k_F} a function

$$\Delta n_{k_F,\exp} = \exp(-U^2 f_{k_F}(t)) + O(U^4) \qquad (6.16)$$

is introduced. In the above formula the logarithmic behavior leads to a power law decay. Thus this approach would reproduce the qualitatively correct behavior in one dimension. The resulting function in two dimensions is shown in Fig. 6.11 as dashed-dotted curve. The initial slope of the curve is retained, but no negative values occur.

However, the 9-loop calculation reveals a jump which lies below the one predicted by this function. This can be attributed to real relaxation effects.
The decrease is captured by the fit function

$$\Delta n_{\text{weak}} = \Delta n_{k_F,\exp}\exp(-\sqrt[4]{(at)^4 + b^4} + b) \qquad (6.17)$$

so that only the deviations from Eq. 6.16 are ascribed to relaxation effects. The leading orders in U^2 and t^2 are known from Eqs. 6.14b and 5.6. These do not describe relaxation. Thus the corrections captured by the exponential function have to be $O(U^4)$ and $O(t^4)$ [46, 157, 158]. Since $\Delta n_{F_k,\exp}$ already reproduces the $t^2 U^2$ result the quartic dependence under the root is chosen to ensure that all corrections describing relaxation effects are $O(U^4)$ and $O(t^4)$.

There have been other studies concerning relaxation. But in contrast to these the study presented here does not base on the assumption that relaxation is present. In other studies a mixture describing the state of the system is already build in by construction [157, 159]. With Eq. 6.17 the curves in the weak quench regime can be fitted. Corresponding fits for some exemplary values of U are depicted in Fig. 6.11. The fitting parameters are included in Fig. 6.8.

Two-dimensional Model

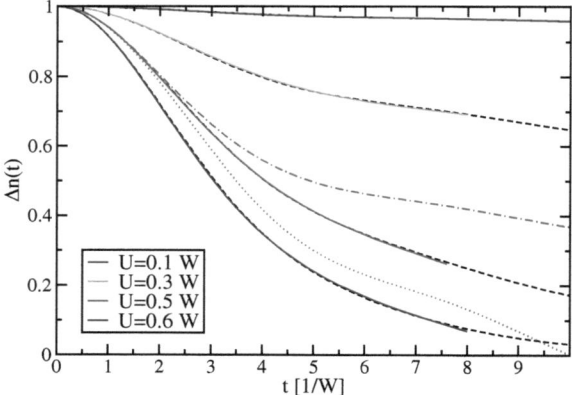

Fig. 6.11.: Jump $\Delta n(t)$ in dependence on the time t. Dashed lines: Fits to the data according to Eq. 6.17. Dotted line: Second order result $\Delta n_{k_F,\text{2nd}}$ as given in (6.14b). Dashed-dotted line: Jump as calculated by (6.16).

6.1.5.2. Relaxation

The fitting parameters for the two regimes fit well to each other (see Fig. 6.8). For the data between $U = 0.65W$ and $U = 0.7W$ it cannot be decided whether the curves belong to the weak or the strong quench regime. A unique fit is not possible within this range and so these values are left out. The two values $U = 0.65W$ and $U = 0.7W$ are given as vertical lines in Fig. 6.8. Obviously the two regimes are again separated by a dynamical transition, similar to what was observed in the one-dimensional model. Thus the appearance of such a transition is indeed generic for quenched Hubbard models.
For quenches to $U = 0.7W$ the relaxation rate reaches values in the order of the band width. For larger interaction strengths U the relaxation rate is quickly decreased to rather small values. For even stronger quenches it can be seen that the period T approaches the

6.1 Half-Filled Two-Dimensional Model

$T = \frac{2\pi}{U}$ curve governed by local Rabi oscillations. Remarkably the relaxation rate also decreases for strong quenches. This is also attributed to the Rabi oscillations. The linear decay in the compactified interaction $\frac{U}{U_C/2+U}$ accounts for a quadratic dependence on the band width $a \propto \frac{W^2}{U}$. This term is the next-leading order in a Magnus expansion [160] which works as follows: The vanishing of the leading order $\propto W$ is explained by the Rabi oscillations. If these are assumed to be within a reference frame where they appear to be static, the fast time dependence of $\hat{H}$ washes out the effects of the hopping and with this the leading order in W, similar to a rotating wave approximation.

In conclusion, the two-dimensional model shows relaxation beyond oscillatory or power law behavior. The relaxation rate is the largest for values of about $U = 0.7W$ where the dynamical transition occurs. Note that in this range the time where convergence is achieved is drastically increased (see Fig. 6.5). For these interaction strengths strong relaxation occurs. Thus also processes spoiling the convergence relax fast and the range of convergence is increased.

6.1.6. Momentum Dependence of the Jump

In this section the dependence of the jump on the actual position on the Fermi surface is discussed. To study the behavior of the jump for different values of the wavevector $\vec{k}$ the jump is calculated at the edges of the Fermi surface $(\pm\pi, 0), (0, \pm\pi)$ and between them at $(\pm\frac{\pi}{2}, \pm\frac{\pi}{2})$ (see Fig. 6.1). Corresponding results for a quench to $U = 0.25W$ derived in nine loops are depicted in Fig. 6.12.

Obviously the jump calculated at $(\frac{\pi}{2}, \frac{\pi}{2})$ decreases faster than the one calculated at $(0, \pi)$ for the observed times. At about $t = 7.5/W$ the curves cross each other and the $(0, \pi)$-curve decreases faster than the $(\frac{\pi}{2}, \frac{\pi}{2})$-curve. Note that this time lies beyond the range of convergence indicated by the dashed lines. Consequently results obtained for this time have to be treated cautiously. However, the effect for small times is captured exactly by the iterated equation of motion approach.

Two-dimensional Model

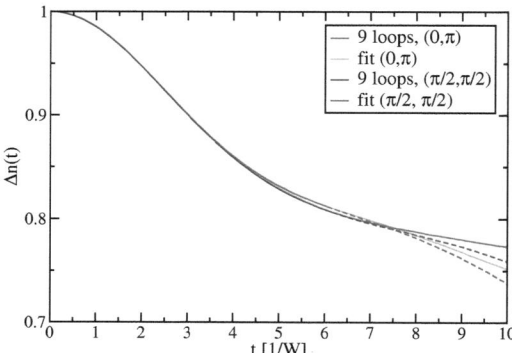

Fig. 6.12.: Results for the jump in a quench with $U = 0.25W$ calculated at $(0, \pi)$ and at $(\frac{\pi}{2}, \frac{\pi}{2})$. Dashed lines: Behavior of the jump for times beyond the range of convergence. Additionally fits based on Eq. 6.17 with the second order result $\Delta n_{k_F, \text{2nd}}$ for the different $\vec{k}$-values are depicted.

Although there is a difference in the curves derived for the different points on the Fermi surface, the effect is rather small for the times captured. For stronger quenches the effect resulting from different positions on the Fermi surface is even smaller as can be seen in Fig. 6.13.

In this figure the jump calculated at $(0, \pi)$ and the one derived for $(\frac{\pi}{2}, \frac{\pi}{2})$ are depicted for various interaction strengths. For $U = 0.5W$ the effect is still visible but fairly small. Again the jump at $(\frac{\pi}{2}, \frac{\pi}{2})$ decreases faster than the one at $(0, \pi)$. Stronger quenches lead to even smaller effects. For $U = 1.0W$ the two curves coincide.

Since these two distinct points on the Fermi surface lead to similar curves for the jump, it can be assumed that the jump behaves similar for all points on the Fermi surface.

In the following the question in how far the effect of different positions on the Fermi surface is captured by the second order calculation presented in Sect. 6.1.4 is addressed. Results for the jump $\Delta n_{k_F, \text{2nd}}$ for a quench to $U = 0.1W$ and for f_{k_F} are shown in Fig. 6.14.

6.1 Half-Filled Two-Dimensional Model

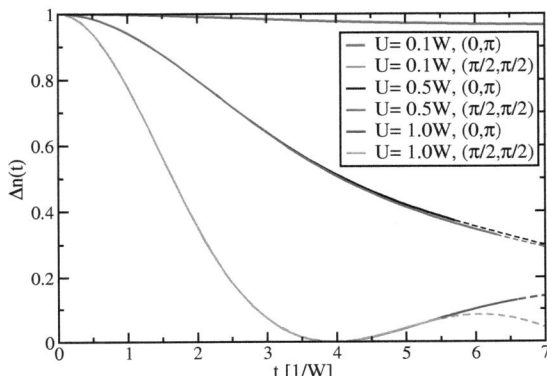

Fig. 6.13.: Jump $\Delta n(t)$ calculated at different positions on the Fermi surface for various interaction strengths U. On increasing U the effect becomes smaller.

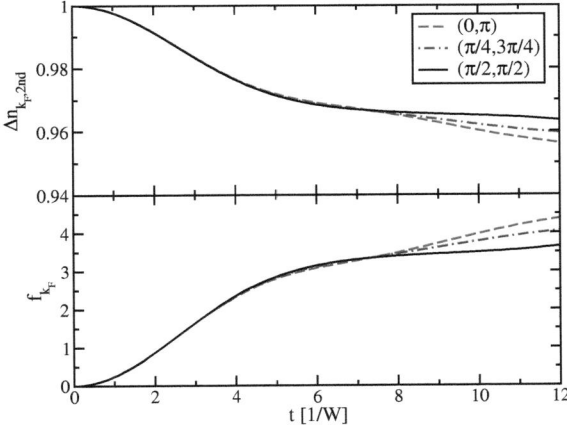

Fig. 6.14.: Upper panel: Results for the jump derived in a second order calculation according to Eq. 6.14a. Lower panel: Result for the function f_{k_F} in Eq. 6.14b as derived at different points on the Fermi surface.

Two-dimensional Model

The second order calculation already shows a dependence of the jump on the position on the Fermi surface. For small times the jump at $(\frac{\pi}{2},\frac{\pi}{2})$ decreases faster than the one at $(\frac{\pi}{4},\frac{3\pi}{4})$ and the one at the edge of the Fermi surface at $(0,\pi)$. As expected the curve calculated for $(\frac{\pi}{4},\frac{3\pi}{4})$ lies in between the other curves. For longer times the curves cross each other and the curve for $(0,\pi)$ shows the fastest decrease. However, the results are strictly U^2-results and have to be treated cautiously.

Recently Tsuji *et al.* calculated the jump by the use of a dynamical cluster approach in combination with iterated perturbation theory (DCA/IPT) [161]. Corresponding curves for a quench to $U = 0.25W$ are shown as dotted curves in Fig. 6.15.

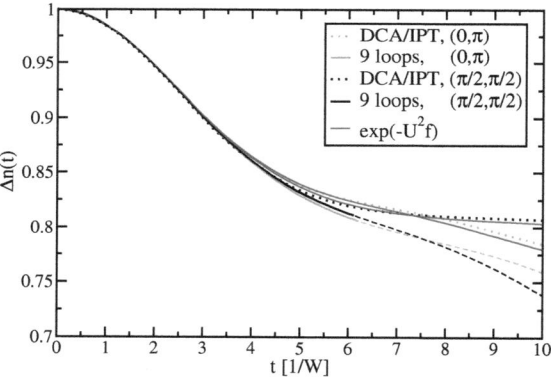

Fig. 6.15.: Results for the jump with $U = 0.25W$ derived in different approaches. Solid lines: Results of the iterated equation of motion approach with nine loops. Dotted lines: Results obtained by DCA/IPT [161]. Green solid lines: Jump derived according to Eq. 6.16.

The curves derived by the DCA/IPT approach agree qualitatively. Concerning numbers these curves differ already for $t \approx 3/W$ from the exact results obtained by the iterated equation of motion approach.

6.1 Half-Filled Two-Dimensional Model

The difference in the two approaches is larger than the difference of the curves derived by the DCA/IPT approach for different points on the Fermi surface. Results obtained in the second order approach based on the function f_{k_F} according to Eq. 6.16 can be performed for different points on the Fermi surface. The corresponding results (green lines in Fig. 6.15) already show a good agreement to the results of the DCA/IPT approach. This is to be attributed to the use of perturbation theory in the DCA/IPT approach.

Based on the second order results for the quench at different points of the Fermi surface fits according to Eq. 6.17 capturing the relaxation of the curves can be produced. Results of the fit for a quench to $U = 0.25W$ are included in Fig. 6.12.

Obviously the fits are very accurate. Some resulting parameters for the jump at $(0,\pi)$ and the jump at $(\frac{\pi}{2},\frac{\pi}{2})$ are depicted in Fig. 6.16. The parameter a denotes the relaxation rate of the exponential decrease. The differences in the fit parameters for different positions on the Fermi surface are small. Consequently there is only a small effect on the relaxation rate derived from these fits to the data of the iterated equation of motion approach.

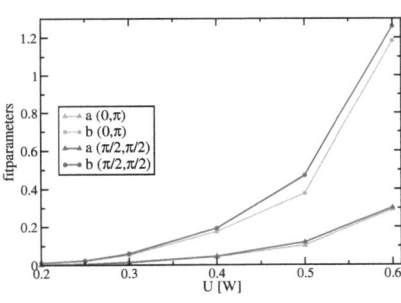

Fig. 6.16.: Parameters used in the fits according to Eq. 6.17 for jumps calculated at $(0,\pi)$ and at $(\frac{\pi}{2},\frac{\pi}{2})$.

In conclusion there is a momentum dependence concerning the behavior of the jump, but the effect is fairly small on the time scales discussed. For stronger quenches this effect gets even smaller.

Two-dimensional Model

6.1.7. Full Momentum Distribution

Due to the vast amount of terms appearing within a commutation and since many points on the Fermi surface have to be evaluated in the two-dimensional model the calculation of the full momentum distribution is restricted to 6-loop calculations. Like in the one-dimensional case the results are adjusted by the Gibbs phenomenon. Results for a rather strong quench to $U = 2.0W$ are given in Fig. 6.17.

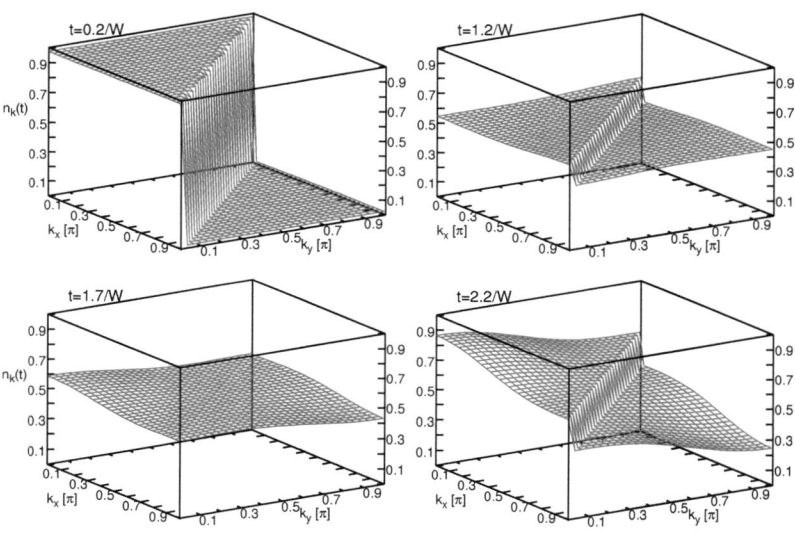

Fig. 6.17.: Full momentum distribution for the half-filled model with $U = 2.0W$ as derived in six loops for various times. Due to point group symmetries it is sufficient to consider one quadrant in the Brillouin zone. After vanishing completely at $t = 1.7/W$ the jump is recovered for larger times.

As point group symmetries apply it is sufficient to calculate the momentum distribution only for one quadrant of the Brillouin zone. From the flat shape around the Fermi surface it can be seen that the jump behaves similar for all these k-values as expected from the

6.1 Half-Filled Two-Dimensional Model

discussion in Sect. 6.1.6. For $t = 0.2/W$ the momentum distribution still shows its box shaped form with a slightly decreased jump. At $t = 1.2/W$ the jump is significantly reduced and a slight curvature of $n_k(t)$ can be observed. For larger times $t = 1.7/W$ the jump vanishes completely leaving behind a rather constant momentum distribution. This signals that the system is in an essentially local state. Remarkably the jump is recovered for larger times. For this large interaction a time of $t = 2.2/W$ is already beyond the range of convergence of a 6-loop calculation. Thus the data in this regime do not show the highest accuracy. It can be seen that the momentum distribution is slightly shifted upwards. But the qualitative behavior is not changed. For large interactions U the momentum distribution shows oscillations stretching over the whole Brillouin zone, strongly reminiscent of collapse and revival behavior as observed experimentally for bosons [8]. The results suggest to perform similar experiments for fermionic systems.

6.2. Doped Two-Dimensional Model

Upon doping the Fermi surface is changed from a boxed shape to a curvy shape losing the particle-hole symmetry. In this chapter the influence of various fillings on the dynamics of the two-dimensional model is studied.

Before the iterated equation of motion approach can be applied the correlations have to be determined.
For the doped model the bounds for the derivation of the correlations $\langle \hat{c}^+_{\vec{r},\sigma} \hat{c}_{0,\sigma} \rangle$ with the vector $\vec{r} = (l,j)^T$ can be determined by the dispersion. It is parametrized by

$$\epsilon_F = -4J\cos(s)\cos(p) \qquad (6.18a)$$

according to

$$s = \pm \arccos\left(-\frac{\epsilon_F}{4J}\frac{1}{\cos(p)}\right). \qquad (6.18b)$$

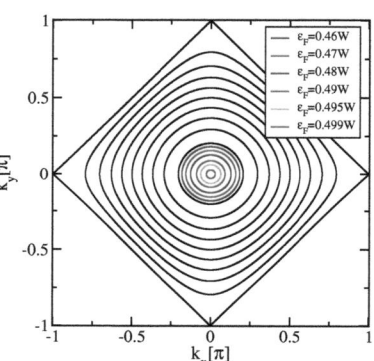

Fig. 6.18.: Fermi surface for various values of ϵ_F. Black lines: the Fermi energy is increased from 0 to 0.45 in steps of 0.05 from the outermost to the innermost line.

The corresponding bounds for p are by definition

$$s = \pm \arccos\underbrace{\left(-\frac{\epsilon_F}{4J}\frac{1}{\cos(p)}\right)}_{\leq 1} \qquad (6.19a)$$

$$\Rightarrow p = \pm \arccos\left(-\frac{\epsilon_F}{4J}\right) \qquad (6.19b)$$

6.2 Doped Two-Dimensional Model

so that the particle number is determined by solving

$$n = \frac{2}{(2\pi)^2} \int_{-\arccos\left(-\frac{\epsilon_F}{4J}\right)}^{\arccos\left(-\frac{\epsilon_F}{4J}\right)} \int_{-\arccos\left(-\frac{\epsilon_F}{4J}\frac{1}{\cos(p)}\right)}^{\arccos\left(-\frac{\epsilon_F}{4J}\frac{1}{\cos(p)}\right)} ds dp \quad (6.20a)$$

$$= \frac{4}{(2\pi)^2} \int_{-\arccos\left(-\frac{\epsilon_F}{4J}\right)}^{\arccos\left(-\frac{\epsilon_F}{4J}\right)} \arccos\left(-\frac{\epsilon_F}{4J}\frac{1}{\cos(p)}\right) dp. \quad (6.20b)$$

The dependence of the particle number n on the Fermi energy ϵ_F is given in Fig. 6.19, with $n = 0$ for $\epsilon_F = -0.5W$. In the other limit $\epsilon_F = 0$ the half-filled case is recovered. For small Fermi energies the curve shows a linear increase as presented in the inset.

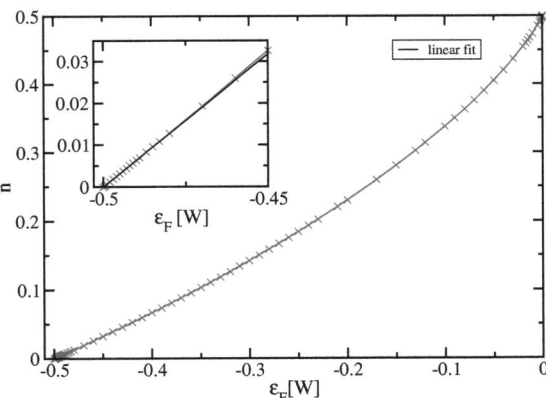

Fig. 6.19.: Dependence of the particle number on the Fermi energy ϵ_F. The inset shows the linear increase observed for $\epsilon_F \approx -0.5W$.

Two-dimensional Model

With these bounds the correlations can be expressed as

$$\langle c^\dagger_{lj,\uparrow} c_{00,\uparrow}\rangle = \frac{1}{(2\pi)^2}\int\int 2\cos((s+p)l+(s-p)j)\mathrm{d}s\mathrm{d}p \quad (6.21\mathrm{a})$$

$$= \frac{1}{(2\pi)^2}\int\int 2\cos((l+j)s)\cos((l-j)p)\mathrm{d}s\mathrm{d}p \quad (6.21\mathrm{b})$$

which is converted to

$$\langle c^\dagger_{lj,\uparrow} c_{00,\uparrow}\rangle = \frac{1}{(2\pi)^2}\int 2\cos(p(l-j))\frac{1}{l+j}\sin(s(l+j))\Big|_{-\arccos\left(-\frac{\epsilon_F}{4J}\frac{1}{\cos(p)}\right)}^{\arccos\left(-\frac{\epsilon_F}{4J}\frac{1}{\cos(p)}\right)} \mathrm{d}p. \quad (6.22)$$

Finally the correlations are given by solving

$$\langle c^\dagger_{lj,\uparrow} c_{00,\uparrow}\rangle = 2\cdot\frac{1}{(2\pi)^2}\int_0^{\arccos\left(-\frac{\epsilon_F}{4t}\right)} \frac{4}{l+j}\cos(p(l-j))$$

$$\cdot\sin\left(\arccos\left(-\frac{\epsilon_F}{4t}\frac{1}{\cos(p)}\right)(l+j)\right)\mathrm{d}p \quad (6.23)$$

numerically. Consequently the calculation of correlations is more demanding than in the half-filled case. With these correlations the jump $\Delta n(t)$ can be determined by the iterated equation of motion approach.

The jump for various values of the filling factor n for one spin species in a quench to $U = 0.7W$ is depicted in Fig. 6.20. For this interaction strength a significant relaxation rate a is observed.

It can be seen that the range of convergence for the half-filled case is rather large compared to the curves for other fillings. From the results for the one-dimensional model it is expected that the range of convergence is decreased on increasing n (for $n < 0.5$).

6.2 Doped Two-Dimensional Model

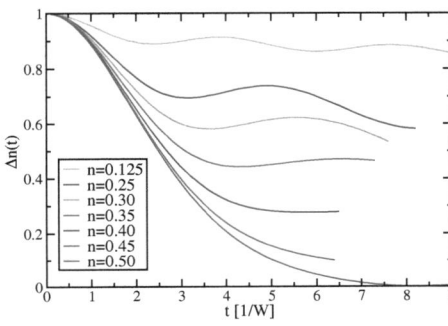

Fig. 6.20.: Jump $\Delta n(t)$ for a quench to $U = 0.7W$ in dependence on time t obtained in a 9-loop calculation. The curves are shown for various fillings, with n denoting the filling factor for one spin species. (n increases from top to bottom.)

This effect can be seen for the curves for $n = 0.125$ to $n = 0.45$. Surprisingly the range of convergence is increased on passing from $n = 0.45$, i.e., 10% of doping, to the half-filled case. This is to be attributed to the fast relaxation in the intermediate U range for half-filling. Presumably the fast relaxation also implies a fast relaxation of processes spoiling the convergence so that their influence is reduced.

Under the influence of doping this fast relaxation is lost and the curves show a decay of the jump superseded by weak oscillations as can be seen for instance for quarter-filling with $n = 0.25$. The fast relaxation is not just shifted towards larger U upon doping. This is shown in Fig. 6.21, where results for the jump in the quarter-filled model are shown as derived in an 8-loop calculation. The plot shows results for quenches with interactions between $U = 0.1W$ and $U = 2.0W$. On increasing U the curves appear to be more and more squeezed with increasing oscillation amplitudes. Thus the curves pass from a gradually decreasing jump to pronounced oscillations like in the half-filled case, but without curves showing fast relaxation with a pure exponential decrease as in the case $U = 0.7W$ at half-filling. Consequently the dynamical transition is lost under the influence of doping.

Just like in the one-dimensional case (see Sect. 5.2) even for large quenches the jump does not exhibit zeros away from half-filling. For $U = 20W$ the curves show oscillations with a fixed period, but the

Two-dimensional Model

minima are finite for $n \neq 0.5$.

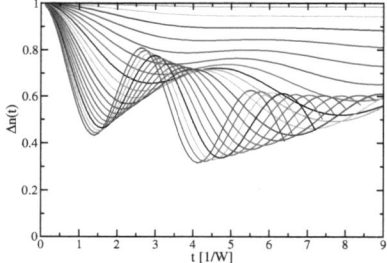
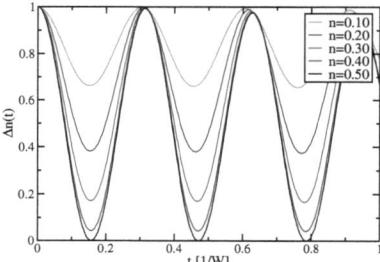

Fig. 6.21.: Jump for the quarter-filled model ($n = 0.25$) derived in eight loops for interaction strengths from $U = 0.1W$ to $U = 2.0W$ in steps of $0.01W$ (from top to botttom).

Fig. 6.22.: Jump for a large quench with $U = 20W$ for various filling factors n for one spin species as derived in eight loops. Upon doping the minima are shifted upwards.

Examples for quenches with $U \gtrsim 1.0W$ at half-filling leading to oscillations which reach zero are already plotted in Fig. 6.5. These zeros lead to a perfectly flat momentum distribution as discussed for $U = 2.0W$. Upon doping the zeros in the jump $\Delta n(t)$ are shifted upwards (see Fig. 6.22). Even strongest quenches do not show oscillations reaching zero away from half-filling.

6.2.1. Second Order Results away from Half-Filling

Analogously to the half-filled case the second order in U result can also be calculated for the doped system. Results for the function f_{k_F} describing the decrease of the jump for various fillings n are presented in Fig. 6.23. The curves are shown up to a time of $t = 10/W$, which is comparable to the ranges of convergence of the results derived by the equation of motion approach. The black line depicts the result for the half-filled case, showing a logarithmic increase. The curve representing the result for 10% of doping ($n = 0.45$) lies close to the curve for half-filling for small times. For

6.2 Doped Two-Dimensional Model

larger times this curve exhibits a weaker increase than the curve representing half-filling. Having determined f_{k_F} the U^2-result can be calculated. Corresponding curves for quenches to $U = 0.3W$ are given in Fig. 6.24. For larger times the curve for $n = 0.5$ shows a much stronger decrease than the curve derived for $n = 0.45$.

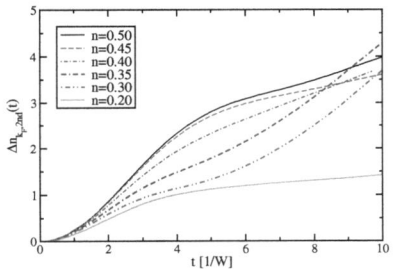
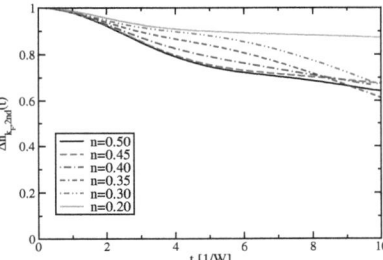

Fig. 6.23.: Second order result f_{k_F} in Eq. 6.14b for the two-dimensional Hubbard model and various fillings in dependence on time t.

Fig. 6.24.: Second order result for various fillings derived from f_{k_F} through Eq. 6.14b as function of time t for a quench to $U = 0.3W$.

Under the influence of doping the Fermi surface changes, which may result in plateaus appearing in the curves for the jump $\Delta n(t)$. Away from half-filling the plateaus are expected to appear at larger times, since the dispersion does not exhibit flat parts. Thus longer observation times are needed to detect the plateaus for larger doping.

The flat part of the $n = 0.45$-curve around $t = 10/W$ could be the beginning of a plateau, but this would be beyond the range of convergence of the full iterated equation of motion approach. Thus the U^2-result does not show prethermalization plateaus on the accessible time scales. For larger doping, for instance for $n = 0.35$, f_{k_F} shows a strong increase, which goes beyond the increase of the curve for half-filling (see Fig. 6.23). Consequently the jump does not exhibit a plateau within this time regime (see Fig. 6.24). A doping with 60%, i.e., $n = 0.20$, leads to a rather flat curve for f_{k_F} and

Two-dimensional Model

a slowly decreasing jump in Fig. 6.24. However, the jump still decreases and there is no plateau visible up to $t = 10/W$. To decide whether plateaus exist beyond this time further work is called for. Coming back to the case with $n = 0.45$, results for the jump derived in the U^2-calculation are opposed to the corresponding results derived in a 9-loop calculation with $U = 0.3W$ in Fig. 6.25. It can be seen that the U^2-result lies below the exact 9-loop result. Just like in the half-filled case an exponential approach given by Eq. 6.16 results in a curve lying above the other curves. Again this result can be used to fit the exact result in the range where the curve is converged.

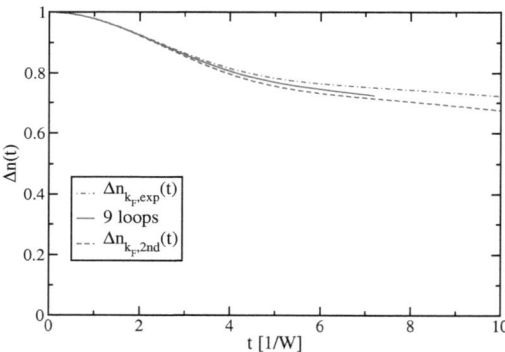

Fig. 6.25.: Jump $\Delta n(t)$ for a quench to $U = 0.3W$ and 10% of doping leading to $n = 0.45$. Solid line: Result of the iterated equation of motion approach with nine loops. Dashed line: Second order result $\Delta n_{k_F,2nd}(t)$ Dashed-dotted line: Jump derived by Eq. 6.16.

With this technique the parameters a and b can be fitted - as in the half-filled case. But from the results for quenches to $U = 20W$ and various fillings (Fig. 6.22) it can already be concluded that there is no dynamical transition away from half-filling. Even for very large interaction strengths U the jump does not exhibit oscillations touching zero so that a transition from a gradually decreasing jump to a jump showing this type of oscillations ceases to exist.

6.2 Doped Two-Dimensional Model

6.2.2. Momentum Distribution

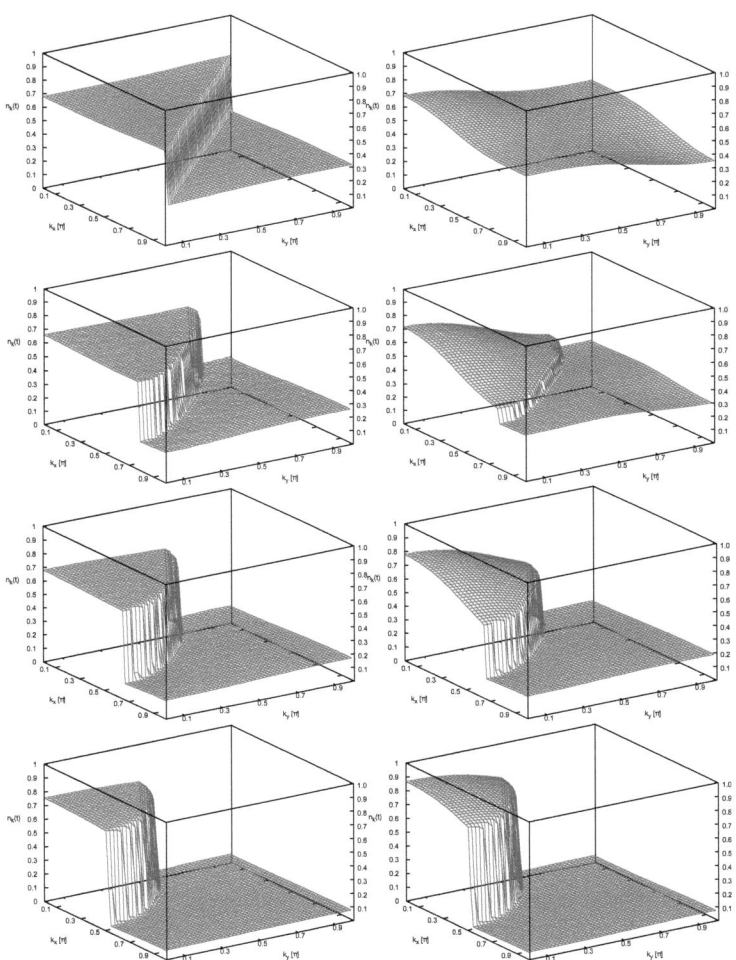

Fig. 6.26.: Momentum distribution for a quench to $U = 1.5W$ derived in 6-loop calculations. The curves are shown for various filling factors n with $n = 0.5, 0.4, 0.3, 0.2$ from top to bottom. Left column: Momentum distribution for $t = 1.2/W$. Right column: Momentum distribution for $t = 2.2/W$.

Two-dimensional Model

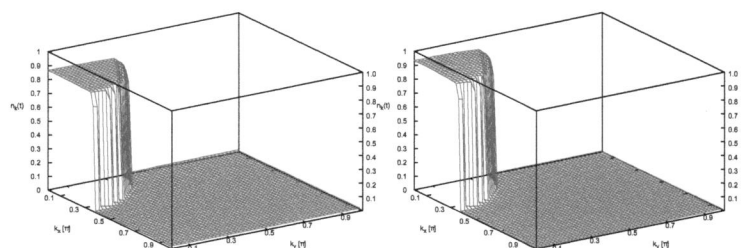

Fig. 6.27.: Momentum distribution for a quench to $U = 1.5W$ and a filling factor of $n = 0.1$ for $t = 1.2/W$ (left panel) and $t = 2.2/W$ (right panel)

The behavior of the complete momentum distribution for $U = 1.5W$ is depicted in Figs. 6.26 and 6.27 for filling factors of $n = 0.5, 0.4, 0.3, 0.2$ and 0.1 from top to bottom. On the left hand side the momentum distributions are depicted for a time $t = 1.2W$. At this time the jump is still nonzero for half-filling, see Fig. 6.5. But due to the large interaction strength the jump is already significantly reduced for $n = 0.5$. On decreasing fillings the jump becomes larger. Besides, the jump is shifted and the shape of the Fermi surface is changed. For 80% of doping, i.e., $n = 0.1$ the jump is close to $\Delta n = 0.9$. On the right hand side the corresponding momentum distributions are shown for $t = 2.2W$. For this time the jump in the half-filled case reaches zero. Consequently the momentum distribution is flat with a slight curvature.

Upon doping the flatness is lost and it can be observed how the jump is built up on doping the system away from half-filling. For $n = 0.4$ a small jump is exhibited which interrupts the curvature of the momentum distribution. Larger doping leads to a pronounced jump and flatter curves to the left of the jump. For smallest fillings n the momentum distribution shows only weak changes from $t = 1.2/W$ to $t = 2.2/W$ as can be seen in Fig. 6.27. The momentum distribution for the half-filled case shows a significant change for the two times, whereas the corresponding results for $n = 0.1$ depict a momentum distribution with weak oscillations in the depth of the jump.

Concluding the completely flat stretched momentum distribution observed at half-filling for $t = 2.2/W$ changes upon doping to a momentum distribution with a more pronounced jump at the Fermi surface.

7. Summary and Outlook

7.1. Summary

In the present thesis the dynamics of Hubbard models after interaction quenches is studied. For the description of these models a well controlled semi-analytical method based on the expansion of the equations of motion is developed. The method yields exact results in the regime where convergence is achieved but the observation times are limited by the computational effort. The reliability of the results goes far beyond the second order results presented in Sect. 4.3 and Sect. 6.1.5. Especially for large interaction strengths $U \gtrsim W$ the method is very powerful. In this regime pronounced oscillations governed by local Rabi oscillations between a singly occupied state and a state containing a particle and a particle-hole pair, as discussed in Sect. 5.4.1, occur. The iterated equation of motion method is very flexible and allows to study the influence of doping on the time evolution of the system (see Sect. 5.2).

Additionally to the results of the equation of motion approach the jump in the momentum distribution is discussed by the use of bosonization techniques in Sect. 5.3. As test for the validity of the bosonization results a model of spinless fermions is discussed in a first step. In equilibrium this model is analytically solvable by Bethe-ansatz and yields exponents describing the time dependence of the jump after the quench fairly well. In contrast to this the exponents of the Hubbard model have to be determined numerically (see App. C). Surprisingly the behavior of the jump can only partially be understood by bosonization techniques. On the accessible time scales the jump indeed shows a power law decrease, but the exponents differ from the ones expected from bosonization theory. Instead a bosonization around the Fermi sea, as presented in Sect. 5.3.4, yields appropriate exponents. This is

Summary and Outlook

explained by the large energies implied by the quench.

Exposed to the quench the Hubbard model exhibits a dynamical transition as indicated by an anomaly in the period of the oscillations occuring in the time evolution of the jump discussed in Sect. 5.4.2. The transition separates the regime of weak quenches from the one describing strong quenches. The strong quench regime is distinct by the existence of zeros in the oscillations. Upon doping this feature is lost as shown in Sect. 5.2. Surprisingly even for smallest changes in the filling the zeros in the oscillations vanish completely even for very strong quenches ($U = 100W$).

In Sect. 6.1.2 it is shown that this dynamical transition also occurs for the two-dimensional model. As these very distinct models show a dynamical transition and this transition was also found for an infinite-dimensional Hubbard model, it is assumed that this is a generic feature of quenched Hubbard models.

In the last part of this thesis the two-dimensional Hubbard model is studied. In contrast to the one-dimensional model this model is not integrable and indeed it shows a true relaxation. This relaxation can be described by exponentially decreasing functions as given in Sect. 6.1.2. In Sect. 6.1.7 the full momentum distribution for the two-dimensional model is discussed. It shows collapse and revival oscillations like the ones observed in bosonic models. For intermediate times the momentum distribution is completely flat, before the jump is recovered at the Fermi surface. For the times discussed in this thesis there are only small differences in the results for the jump calculated at different points of the Fermi surface.

Even though the two-dimensional model is not integrable, calculations up to second order in U do not show prethermalization plateaus as can be observed in Sect. 6.1.5. This can be understood by the shape of the Fermi surface leading to a quasi one-dimensional behavior. The influence of doping on the two-dimensional model is discussed in Sect. 6.2. Away from half-filling the U^2-calculations do not show plateaus for the observable times.

7.2. Outlook

To check whether the U^2-calculations reveal prethermalization plateaus for longer times, further calculations based on integrating the formula given by Moeckel and Kehrein, which was originally introduced to decribe the infinite-dimensional model, are called for. Besides, the method presented in this thesis can be used to study the influence of different temperatures on the relaxation of the system. Alternatively a next-nearest neighbor hopping t' can be introduced to study whether the breakdown of the description by bosonization is due to the slow convergence of the RG flow in the Hubbard model. By the use of t' the slowly flowing parameter g_1 can be tuned away. Such a study could reveal new insight in the role of integrability for the relaxation of the system. The flexibility of the method allows to study also other models and different initial states.

Another route for future studies is the relaxation of current carrying states. As there is no bath coupled to the system the study of real transport processes is beyond the scope of this method. However, a shifted Fermi sea including a current can be used as initial state. In this way the influence of different shifts on the relaxation can be addressed.

Summary and Outlook

Appendix

A. Second Order Calculations

A.1. Calculation of the Commutator

The commutator of the interaction term $\hat{H}_{\text{int}}$ with terms created by the first application of the Liouville superoperator is given through

$$\sum_{p_1,p_2,l} \left[:\hat{c}^\dagger_{p_1+l,\uparrow}\hat{c}_{p_1,\uparrow}::\hat{c}^\dagger_{p_2-l,\downarrow}\hat{c}_{p_2,\downarrow}:,:\hat{c}^\dagger_{k+q,\uparrow}\hat{c}^\dagger_{k_2-q,\downarrow}\hat{c}_{k_2,\downarrow}: \right]$$

$$= \sum_{p_1,p_2,l} \left[:\hat{c}^\dagger_{p_1+l,\uparrow}\hat{c}_{p_1,\uparrow}:,:\hat{c}^\dagger_{k+q,\uparrow}\hat{c}^\dagger_{k_2-q,\downarrow}\hat{c}_{k_2,\downarrow}: \right] :\hat{c}^\dagger_{p_2-l,\downarrow}\hat{c}_{p_2,\downarrow}:$$

$$+ :\hat{c}^\dagger_{p_1+l,\uparrow}\hat{c}_{p_1,\uparrow}: \left[:\hat{c}^\dagger_{p_2-l,\downarrow}\hat{c}_{p_2,\downarrow}:,:\hat{c}^\dagger_{k+q,\uparrow}\hat{c}^\dagger_{k_2-q,\downarrow}\hat{c}_{k_2,\downarrow}: \right] \quad \text{(A.1)}$$

where the first term yields

$$\left[:\hat{c}^\dagger_{p_1+l,\uparrow}\hat{c}_{p_1,\uparrow}:,:\hat{c}^\dagger_{k+q,\uparrow}\hat{c}^\dagger_{k_2-q,\downarrow}\hat{c}_{k_2,\downarrow}: \right]$$
$$= \hat{c}^\dagger_{p_1+l,\uparrow}\hat{c}^\dagger_{k_2-q,\downarrow}\hat{c}_{k_2,\downarrow}\delta_{p_1,k+q}. \quad \text{(A.2)}$$

The second term is calculated in the same manner

$$\left[:\hat{c}^\dagger_{p_2-l,\downarrow}\hat{c}_{p_2,\downarrow}:,:\hat{c}^\dagger_{k+q,\uparrow}\hat{c}^\dagger_{k_2-q,\downarrow}\hat{c}_{k_2,\downarrow}: \right]$$
$$= -\hat{c}^\dagger_{p_2-l,\downarrow}\hat{c}^\dagger_{k+q,\uparrow}\{\hat{c}_{p_2,\downarrow},\hat{c}^\dagger_{k_2-q,\downarrow}\}\hat{c}_{k_2,\downarrow} - \hat{c}^\dagger_{k+q,\uparrow}\hat{c}^\dagger_{k_2-q,\downarrow}\{\hat{c}^\dagger_{p_2-l,\downarrow},\hat{c}_{k_2,\downarrow}\}\hat{c}_{p_2,\downarrow}. \quad \text{(A.3)}$$

Second Order Calculations

The resulting terms are combined with the additional operators from Eq. A.1 and normal ordered resulting in

$$\sum_{p_1,p_2,l} \hat{c}^\dagger_{p_1+l,\uparrow} \hat{c}_{k_2-q,\downarrow} \hat{c}_{k_2,\downarrow} \delta_{p_1,k+q} : \hat{c}^\dagger_{p_2-l,\downarrow} \hat{c}_{p_2,\downarrow} :$$
$$= \sum_{p_1,p_2,l} : \hat{c}^\dagger_{p_1+l,\uparrow} :: \hat{c}_{k_2-q,\downarrow} \hat{c}_{k_2,\downarrow} :: \hat{c}^\dagger_{p_2-l,\downarrow} \hat{c}_{p_2,\downarrow} : \delta_{p_1,k+q} \quad \text{(A.4a)}$$

as the operators containing k and the ones with the p's are already normal ordered among each other.

$$\sum_{p_1,p_2,l} \hat{c}^\dagger_{p_1+l,\uparrow} \hat{c}_{k_2-q,\downarrow} \hat{c}_{k_2,\downarrow} \delta_{p_1,k+q} : \hat{c}^\dagger_{p_2-l,\downarrow} \hat{c}_{p_2,\downarrow} : =$$

$$\sum_{p_1,p_2,l} : \hat{c}^\dagger_{p_1+l,\uparrow} : \delta_{p_1,k+q} \Big(: \hat{c}_{k_2-q,\downarrow} \hat{c}_{k_2,\downarrow} \hat{c}^\dagger_{p_2-l,\downarrow} \hat{c}_{p_2,\downarrow} :$$
$$+ : \hat{c}_{k_2,\downarrow} \hat{c}^\dagger_{p_2-l,\downarrow} : n_{p_2} \delta_{p_2,k_2-q}$$
$$+ : \hat{c}_{k_2-q,\downarrow} \hat{c}_{p_2,\downarrow} : (1-n_{k_2}) \delta_{k_2,p_2-l}$$
$$+ n_{p_2}(1-n_{k_2}) \delta_{k_2,p_2-l} \delta_{p_2,k_2-q} \Big)$$

For results up to second order in U only the last term has to be considered. The other terms belong to newly created monomials appearing in higher order in U.

$$A = \sum_{p_1,p_2,l} : \hat{c}^\dagger_{p_1+l,\uparrow} : \delta_{p_1,k+q} n_{p_2}(1-n_{k_2}) \delta_{k_2,p_2-l} \delta_{k_2-q,p_2} \quad \text{(A.5a)}$$
$$=: \hat{c}^\dagger_{k,\uparrow} : n_{k_2-q}(1-n_{k_2}) \quad \text{(A.5b)}$$

A.1 Calculation of the Commutator

After normal ordering the second term in Eq.A.1 yields

$$\sum_{p_1,p_2,l} :\hat{c}^\dagger_{p_1+l,\uparrow}\hat{c}_{p_1,\uparrow}: \left(-\hat{c}^\dagger_{p_2-l,\downarrow}\hat{c}^\dagger_{k+q,\uparrow}\hat{c}_{k_2,\downarrow}\delta_{p_2,k_2-q}\right.$$
$$\left.-\hat{c}^\dagger_{k+q,\uparrow}\hat{c}^\dagger_{k_2-q,\downarrow}\hat{c}_{p_2,\downarrow}\delta_{p_2-l,k_2}\right)$$
$$=\sum_{p_1,p_2,l} :\hat{c}^\dagger_{p_1+l,\uparrow}\hat{c}_{p_1,\uparrow}::\hat{c}^\dagger_{k+q,\uparrow}:\left(:\hat{c}^\dagger_{p_2-l,\downarrow}\hat{c}_{k_2,\downarrow}:+n_{k_2}\delta_{p_2-l,k_2}\right)\delta_{p_2,k_2-q}$$
$$+:\hat{c}^\dagger_{p_1+l,\uparrow}\hat{c}_{p_1,\uparrow}::\hat{c}^\dagger_{k+q,\uparrow}:\left(-:\hat{c}^\dagger_{k_2-q,\downarrow}\hat{c}_{p_2,\downarrow}:-n_{p_2}\delta_{k_2-q,p_2}\right)\delta_{p_2-l,k_2}.$$

Again only the terms with a single creation operator have to be considered such as the terms

$$\sum_{p_1,p_2,l} :\hat{c}^\dagger_{p_1+l,\uparrow}\hat{c}_{p_1,\uparrow}::\hat{c}^\dagger_{k+q,\uparrow}:\left(n_{k_2}\delta_{p_2-l,k_2}\delta_{p_2,k_2-q}-n_{p_2}\delta_{k_2-q,p_2}\delta_{p_2-l,k_2}\right)$$

$$=\sum_{l} :\hat{c}^\dagger_{k+q+l,\uparrow}:(1-n_{k+q})\left[n_{k_2}\delta_{k_2+l,k_2-q}-n_{k_2-q}\underbrace{\delta_{k_2-q-l,k_2}}_{\Rightarrow l=-q}\right] \quad \text{(A.6a)}$$

$$=\hat{c}^\dagger_{k,\uparrow}:(1-n_{k+q})(n_{k_2}-n_{k_2-q}). \quad \text{(A.6b)}$$

Combining both terms leads to

$$:\hat{c}^\dagger_{k,\uparrow}:\left(n_{k_2-q}(1-n_{k_2})+(1-n_{k-q})(n_{k_2}-n_{k_2-q})\right)$$

$$=::\hat{c}^\dagger_{k,\uparrow}:\underbrace{\left[-n_{k_2-q}n_{k_2}+n_{k+q}n_{k_2}+n_{k+q}n_{k_2-q}\right]}_{h_{k_2,k,q}}. \quad \text{(A.7)}$$

At zero temperature $T=0$ $n_k \in 0,1$ holds. Thus the term $h_{k_2,k,q}$ is equal to one only if $n_{k_2}=1$ and $n_{k_2-q}=0=n_{k+q}$ or if $n_{k_2-q}=1=n_{k+q}$ and $n_{k_2}=0$. In all other cases the term vanishes.
The corresponding energy difference used to describe the transla-

Second Order Calculations

tional part of the time evolution analogously to Eq. 4.30 is denoted by $d_{k_2,k,q} = \epsilon_{k+q} + \epsilon_{k_2-q} - \epsilon_{k_2}$. Its sign can be deduced by the energies in the two non-vanishing cases. In the first case with $n_{k_2} = 1$ and $n_{k_2-q} = 0 = n_{k+q}$ the energies have to obey $\epsilon_{k_2} < 0, \epsilon_{k_2-q} > 0$ and $\epsilon_{k+q} > 0$ and thus $d_{k_2,k,q} > 0$. The second case with $n_{k_2-q} = 1 = n_{k+q}$ and $n_{k_2} = 0$ fulfills $\epsilon_{k_2} > 0, \epsilon_{k_2-q} < 0$ and $\epsilon_{k+q} < 0$ so that $d_{k_2,k,q} < 0$.

A.2. Determination of the Spectral Density

The spectral density is split into $\rho_G(x) = \rho_+(x) + \rho_-(x)$. The first part obeys $\rho_+(x) > 0$ for $x > 0$ and $\rho_+(x) = 0$ for $x < 0$ and the second part obeys $\rho_-(x) > 0$ for $x < 0$ and $\rho_-(x) = 0$ for $x > 0$. With the definition of $d_{k_2,k_F,q}$ the first part yields

$$\rho_+(x) = \frac{1}{(2\pi)^2} \int_{-\pi}^{\pi}\int_{-\pi}^{\pi} dk_2 dq\, \delta(x - \epsilon_{k_F+q} + \epsilon_{k_2-q} - \epsilon_{k_2}) h_{k_2,k_F,q} \quad \text{(A.8)}$$

where $h_{k_2,k_F,q}$ is nonzero only in two cases. The first case is described by $n_{k_2} = 1, n_{k_2-q} = 0$ and $n_{k_F+q} = 0$ which implies $\epsilon_{k_2} < 0, \epsilon_{k_2-q} > 0$ and $\epsilon_{k_F+q} > 0$ and $d_{k_2,k_F,q} > 0$. Analogously the second case yields $n_{k_2} = 0, n_{k_2-q} = 1, n_{k_F+q} = 1$ and $\epsilon_{k_2} > 0, \epsilon_{k_2-q} < 0, \epsilon_{k_F+q} < 0$ for the energies. In this case the energies imply $d_{k_2,k_F,q} < 0$. The density $\rho_+(x)$ is only nonzero for $x > 0$. Due to the δ-function $x = -d_{k_2,k_F,q}$ holds. Thus the first case leads to $x < 0$ and cannot be fulfilled. Consequently the energies have to be chosen according to the second case. In this case the density reads

$$\rho_+(x) = \frac{1}{(2\pi)^2} \int_{-\pi}^{\pi}\int_{-\pi}^{\pi} dk_2 dq\, \delta(x + \epsilon_{k_F+q} + \epsilon_{k_2-q} - \epsilon_{k_2})$$
$$\cdot \Theta(\epsilon_{k_2})\Theta(-\epsilon_{k_2-q})\Theta(-\epsilon_{k_F+q}). \quad \text{(A.9)}$$

A.2 Determination of the Spectral Density

In the same manner the density

$$\rho_-(x) = \frac{1}{(2\pi)^2} \int_{-\pi}^{\pi}\int_{-\pi}^{\pi} dk_2 dq\, \delta(x + \epsilon_{k_F+q} + \epsilon_{k_2-q} - \epsilon_{k_2})$$
$$\cdot \Theta(-\epsilon_{k_2})\Theta(\epsilon_{k_2-q})\Theta(\epsilon_{k_F+q}) \qquad (A.10)$$

for $x < 0$ can be calculated.
Due to the Heaviside function $\Theta(-\epsilon_{k_2})$ the energy ϵ_{k_2} has to be negative. Thus the range for k_2 is set to $-k_F \leq k_2 \leq k_F$.
With the substitution $q \to k_2 + \tilde{q} + \pi$ the dispersion, $\epsilon_{k_F+q} > 0$ is translated into $-\epsilon_{k_F+q+k_2} > 0$. This implies $-k_F \leq k_F + k_2 + q \leq k_F$ and for half-filling $-\pi \leq k_2 + q \leq 0$. Consequently the integration range for q is set to $-k_F \leq q \leq -|k_2|$ by the Heaviside functions. Satisfying the new integration boundaries the density reads

$$\rho_-(x) = \frac{1}{2\pi^2} \int_{-\frac{\pi}{2}}^{\frac{\pi}{2}} dk_2 \int_{-\frac{\pi}{2}}^{-|k_2|} dq\, \delta(x - \epsilon_{\frac{\pi}{2}+k_2+q} - \epsilon_{-q} - \epsilon_{k_2}) \qquad (A.11)$$

with the bandwidth $W = 4J$ and the energies $\epsilon_k = -\frac{W}{2}\cos(k)$. Inserting this relation into Eq. A.11 yields

$$\rho_-(x) = \frac{1}{\pi^2 W} \int_{-\frac{\pi}{2}}^{\frac{\pi}{2}} dk_2 \int_{-\frac{\pi}{2}}^{-|k_2|} dq\, \delta\left(\frac{2x}{W} + \cos(k_2) + \cos(-q) + \cos(k_2 + q + \frac{\pi}{2})\right)$$

(A.12)

where the argument of the δ-function can be simplified by $\cos(x + \frac{\pi}{2}) = -\sin(x)$ to
$\frac{2x}{W} + \cos(k_2) + \cos(q) - \sin(k_2)\cos(q) - \cos(k_2)\sin(q)$. Splitting the inte-

gral into two parts yields

$$\rho_-(x) = \frac{1}{\pi^2 W} \int_{-\frac{\pi}{2}}^{\frac{\pi}{2}} dk_2 \int_{-\frac{\pi}{2}}^{-|k_2|} dq \, \delta\left(\frac{2x}{W} + \cos(k_2) + \cos(q) - \sin(k_2)\cos(q)\right.$$
$$\left. - \cos(k_2)\sin(q)\right) \quad \text{(A.13)}$$

which leads to

$$\rho_-(x) = \frac{1}{\pi^2 W} \int_0^{\frac{\pi}{2}} dk_2 \int_{-\frac{\pi}{2}}^{-|k_2|} dq \, \delta\left(\frac{2x}{W} + \cos(k_2) + \cos(q)\right.$$
$$\left. - \sqrt{1-\cos(k_2)^2}\cos(q) - \cos(k_2)\sqrt{1-\cos(q)^2}\right) \quad \text{(A.14)}$$
$$+ \delta\left(\frac{2x}{W} + \cos(k_2) + \cos(q) + \sqrt{1-\cos(k_2)^2}\cos(q)\right.$$
$$\left. - \cos(k_2)\sqrt{1-\cos(q)^2}\right).$$

For further calculations it is useful to take advantage of the δ-function and rewrite the density as

$$\rho_-(x) = \frac{1}{\pi^2 W} \int_0^{\frac{\pi}{2}} dk \int_{-\frac{\pi}{2}}^{-|k|} dq \, \delta(f(q)) \quad \text{(A.15a)}$$

$$= \frac{1}{\pi^2 W} \int_0^{\frac{\pi}{2}} \sum_{i=1}^n \frac{dk}{|f'(q_i)|} \quad \text{(A.15b)}$$

with q_i denoting the simple zeros of $f(q)$. Thus the calculation is reduced to finding the simple zeros of the argument

A.2 Determination of the Spectral Density

of the δ-functions. For the first δ-function the zeros of $f(q) = \frac{2x}{W} + \cos(k) + \cos(q) - \sqrt{1-\cos(k)^2}\cos(q) - \cos(k)\sqrt{1-\cos(q)^2}$ in the interval $[-\frac{\pi}{2}, -|k|]$ are needed. Here and in the following the index of k_2 is omitted.

The zeros are given by

$$\cos(\tilde{q}) = -\frac{1}{2}\left(\frac{2x}{W} + \cos(k)\right) \pm \sqrt{\frac{1}{4}\left(\frac{2x}{W} + \cos(k)\right)^2 - \frac{2x}{W}\frac{\frac{4x}{W} + \cos(k)}{1 - \sqrt{1-\cos(k)^2}}}. \quad (A.16)$$

In the following it has to be checked whether both signs in Eq. A.16 are solutions of $f(q) = 0$ or a spurious solution appeared by squaring the equation.

For simplicity $\cos(q)$ and $\cos(k)$ are substituted by $\cos(q) = v$ and $\cos(k) = u$. Then the first derivative simplifies to

$$f'(v) = 1 - \sqrt{1-u^2} + \frac{uv}{\sqrt{1-v^2}} \quad (A.17)$$

and the integral is given through

$$\rho_-(x) = \frac{1}{\pi^2 W} \int_0^{\frac{\pi}{2}} dk \int_{-1}^{-\cos(k)} dv \frac{1}{\sqrt{1-v^2}} \delta(f(v)) \quad (A.18a)$$

$$= \frac{1}{\pi^2 W} \int_{lB}^{uB} du \frac{1}{\sqrt{1-u^2}} \sum_i \frac{1}{\sqrt{1-v_i}} \frac{1}{|1 - \sqrt{1-u^2} + \frac{uv_i}{\sqrt{1-v_i^2}}|} \quad (A.18b)$$

where v_i denotes the simple zeros of $f(v)$. The boundaries have to be chosen so that the solutions v_i are real.

The upper boundary u_B is given by the condition $u_B = v_i$ and the lower boundary by $v_i = 0$. Then the integral is solved numerically. With $\rho_-(x)$ and $\rho_+(x)$ the Green function $\mathcal{G}(\omega)$ can be calculated by

Second Order Calculations

Eq. 4.53.

B. Other Truncation Schemes

In this section different truncation schemes for the equation of motion approach are presented. These schemes are either much more demanding than the one used in this thesis while obtaining the same ranges of convergence, or they are only valid for special cases. Due to the drawbacks of these approaches they are not used in this thesis. For simplicity the results are shown for the one-dimensional model at half-filling.

B.1. Truncation According to Order in U and J

In the first truncation scheme the terms to be considered are chosen according to the number of commutations with the interaction part or the hopping term of the Hamiltonian in which they appear. For small interaction strengths U it can be expected that only a few commutations with the interaction term are necessary to describe the dynamics of the system over a rather long time. Thus this case accounts for a truncation scheme according to the number of commutations with the interaction term $\hat{H}_{\text{int}}$. Results of such a truncation scheme for $U = 0.2W$ can be found in Fig. B.1. The number of commutations with the interaction term is denoted by o. The convergence of the results is quickly increased on increasing o. From the inset it can be deduced that six commutations with the interaction term (while performing 11 commutations with the hopping term) are sufficient to describe the dynamics over fairly long times. Such a calculation includes 27701 monomials and is thus comparable to a 9-loop calculation. Consequently this truncation scheme is favorable for small interactions U.

On the contrary larger quenches cannot be described by this approach. Results for a quench to $U = 2.0W$ are given in Fig. B.2. The convergence is increased rather quickly for

Other Truncation Schemes

two and three commutations. However, for more commutations nearly no progress is made. Comparing the $o = 5$ and the $o = 6$ curve to each other, the results are reliable up to $t \approx 2.0/W$.

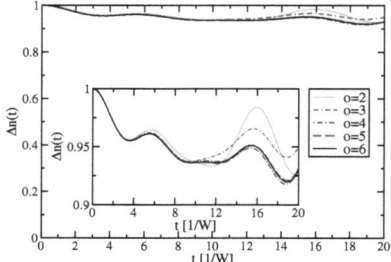 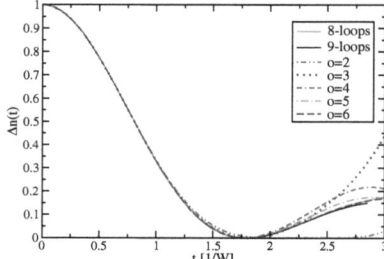

Fig. B.1.: Jump for $U = 0.2W$ obtained in calculations based on a truncation according to the number of commutations with the interaction term o. In the inset the good convergence on increasing o can be observed.

Fig. B.2.: Results for a larger quench $U = 2.0W$ for the truncation scheme based on the number of commutations with the interaction o. As reference curves for calculations with 8 and 9 loops are included.

The $o = 6$ curve lies above the curve obtained in a full 8-loop calculation. The 8-loop calculation includes only 7851 monomials, which is much less than in the $o = 6$ calculation. With the 9-loop curve as reference the 8-loop calculation is valid up to $t \approx 2.7/W$. Consequently the $o = 6$ calculation is much more demanding but leads to a weaker convergence. For large U this truncation scheme should not be applied. On the other hand a truncation scheme according to the number of commutations with the hopping term leads to improvements in this range of U (not shown for brevity).

However, these approaches are not useful in the intermediate range $U \approx W$, where the dynamical transition takes place.

B.2. Omin-Omax Truncation

The next truncation scheme is intended to reproduce the local expectation value $\langle \hat{n}_0 \rangle(t)$ correctly up to a given order N in time t. To achieve a correct result up to a given order N a maximal order O_{max} and a minimal order O_{min} are introduced for each monomial appearing in the calculation. The minimal order O_{min} is the minimal order in t in which the corresponding monomial appears. Due to the structure of the differential equations and the initial conditions this order is given by the number of commutations performed to create the considered monomial for the first time. The maximal order O_{max} denotes the order up to which the prefactor of the monomial has to be known to reproduce results for the expectation value correct up to t^N. This order is determined by the orders of the other monomials with which the term is combined on evaluating $\langle \hat{n}_0 \rangle(t)$. If the orders O_{min} and O_{max} are equal for a given monomial, this monomial and all differential equation entries corresponding to this monomial can be neglected. In this way a much sparser system of differential equations is obtained. Exemplary results for a jump to $U = 1.0W$ are given in Figs. B.3 and B.4.

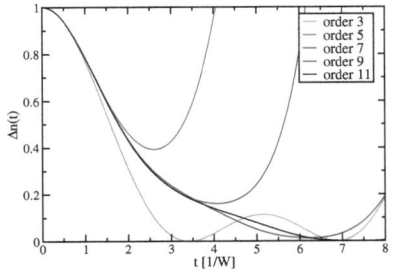
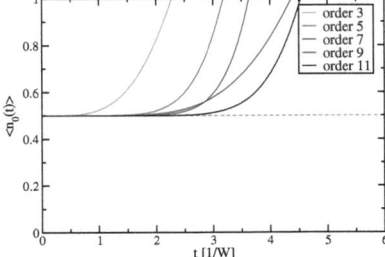

Fig. B.3.: Jump $\Delta n(t)$ for a quench to $U = 1.0W$ as derived by the truncation according to the orders Omin and Omax for different orders.

Fig. B.4.: Local expectation value measuring the convergence of the calculations for a quench to $U = 1.0W$ for different orders.

The convergence of the results is improved on increasing order. In

Other Truncation Schemes

the same sense the convergence of the local expectation value is increased on increasing order. Consequently the truncation works well, as higher orders lead to more accurate results. A comparison of results for different orders can be used to measure the convergence of each calculation.

Besides the increase in range of convergence with increasing orders, it has to be checked if the convergence is better than in a comparable full calculation. For a quench to $U = 2.0W$ such a comparison is given in Fig. B.5.

The results of the order 10 and the order-11 calculation coincide up to $t \approx 2.0/W$. In contrast to this the full 8-loop calculation and the full 9-loop calculation coincide up to larger times $t \approx 2.8/W$. Thus longer ranges of convergence are reached with less commutations with the loop truncation.

A comparison of the results for the local expectation value (Fig. B.6) shows that the convergence concerning this observable is even worse for the Omin-Omax truncation.

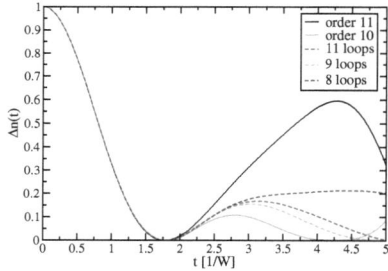

Fig. B.5.: Jump for a quench with $U = 2.0W$ derived in the Omin-Omax truncation in different orders compared to the results of calculations with different loop numbers.

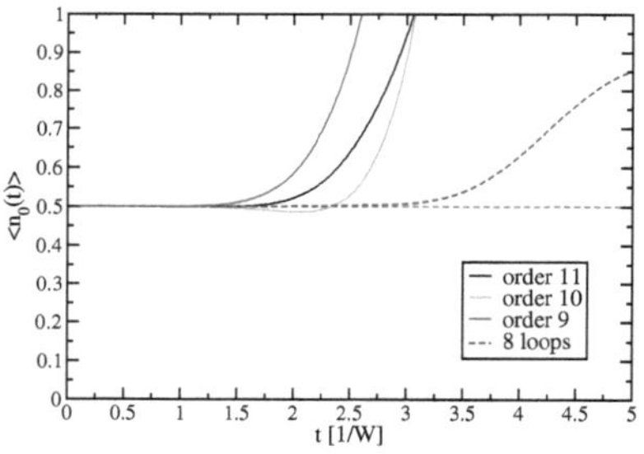

Fig. B.6.: Local expectation value for $U = 2.0W$ as derived in the Omin-Omax truncation with different orders compared to the results of the 8-loop calculation.

The results for a calculation with order 11 is valid up to $t \approx 1.7/W$, whereas the full 8-loop calculation is reliable until $t \approx 2.8/W$. Consequently the self-consistent loop truncation yields better results than

B.2 Omin-Omax Truncation

the Omin-Omax truncation. This is explained by the numerical solution of the differential equation. As no expansion in time is applied, but a Runge-Kutta algorithm, also higher orders in t are created in a calculation with a finite number of loops m.

Other Truncation Schemes

C. Determination of the Luttinger Parameters

In this appendix the derivation of the Luttinger parameters for the one-dimensional Hubbard model is explained. The Luttinger parameters should be determined for vanishing external field $B = 0$. In the Bethe ansatz [129] the integral boundaries $-Q, Q$ and $-A, A$ are given as functions of the chemical potential μ and the external field B. The boundaries lie in the range $0 < Q \leq \pi$ and $0 < A \leq \infty$. As the magnetization is zero for a vanishing field B, the boundary A is set to $A = \infty$ in this case.

The second boundary Q determines the electron density and has to be calculated iteratively. Thus a value for Q in the given range is chosen and proceeded as explained in the following.

1. In a first step, the density function $\rho(k)$ has to be determined. In the Bethe ansatz this function is given by [129]

$$\rho(k) = \frac{1}{2\pi} + \int_{Q}^{-Q} dk' \cos(k) R(\sin(k') - \sin(k)) \rho(k') \quad \text{with } |k| \leq Q \tag{C.1}$$

which is a Fredholm equation of the second kind [162] with the so-called kernel R defined by

$$R(\sin(k') - \sin(k)) = \int_{-\infty}^{\infty} \frac{d\omega}{2\pi} \frac{e^{i\omega(\sin(k') - \sin(k))}}{1 + e^{2U|\omega|}}. \tag{C.2}$$

As the function $e^{2U|\omega|}$ is an even function of ω, the integral is

Determination of the Luttinger Parameters

simplified to

$$R(\sin(k') - \sin(k)) = \int_0^\infty \frac{d\omega}{\pi} \frac{\cos(\omega(\sin(k') - \sin(k)))}{1 + e^{2U\omega}}. \quad \text{(C.3)}$$

The density function $\rho(k)$ determines the particle number n via

$$\int_{-Q}^{Q} \rho(k) dk = n. \quad \text{(C.4)}$$

with the boundaries Q and $-Q$. Consequently Eq. C.1 is a coupled integral equation where the boundaries Q are determined by the solution itself. One way to solve this Fredholm equation of the second kind is the *Nystrom* method [162]. The general form of such a Fredholm equation reads

$$f(t) = \int_a^b K(t,s) f(s) ds + g(t) \quad \text{(C.5)}$$

with the kernel $K(t,s)$. Before applying the Nystrom method a quadrature rule

$$\int_a^b y(s) ds = \sum_{j=0}^{N-1} w_j y(s_j) \quad \text{(C.6)}$$

with the weights w_j and the corresponding abscissas s_j has to be chosen. In the following the Gauss-Legendre quadrature is used to determine the weights and abscissas in

$$f(t) = \sum_{j=0}^{N-1} w_j K(t, s_j) f(s_j) + g(t) \quad \text{(C.7)}$$

with the function $g(t) = \frac{1}{2\pi}$ in this case. The kernel $K(t,s)(x)$ is given through Eq. C.3 as

$$K(t,s)(x) = \int_0^\infty \frac{1}{\pi} \frac{\cos(\omega(\sin(s)-\sin(t)))}{1+e^2 U\omega} d\omega. \tag{C.8}$$

In a first step the equation is evaluated at the quadrature points t_i with the vectors $f_i = f(t_i)$ and g_i analogously. The kernel $K(t_i, s_j)$ can then be rewritten as matrix $K_{i,j}$ and weighted according to $\tilde{K}_{i,j} = K_{i,j} w_j$ leading to the matrix equation

$$(1 - \lambda \tilde{K})\vec{f} = \vec{g} \tag{C.9}$$

with $\lambda = 1$ in the following.

Having rewritten Eq. C.5 in matrix form it can be solved by the usual matrix methods such as the LU-decomposition.

2. Once the density ρ is calculated at the quadrature points, the interpolation is used to evaluate the integral

$$\int_{-Q}^{Q} \rho(k) dk = n. \tag{C.10}$$

and to determine the electron density n.

At this point it is checked whether the required particle number n is recovered. If the density does not fit, a different value for Q is chosen and the steps 1) and 2) are repeated until the wanted density is achieved.

Here the dependence of the density n on the boundary Q helps to find the right value. The density is a linear increasing function of Q with a maximum of $n = 1$ for the half-filled case, reached at $Q = \pi$.

Determination of the Luttinger Parameters

3. In the next step, the function $\kappa(k)$ [129] is determined by

$$\kappa(k) = -2\cos(k) - \mu - 2U + \int_{-Q}^{Q} dk' \cos(k') R(\sin(k') - \sin(k)) \kappa(k') \tag{C.11}$$

with R given in Eq. C.2. In the following the derivative of $\kappa(k)$

$$\kappa'(k) = 2\sin(k) + \frac{\partial}{\partial k} \int_{-Q}^{Q} dk' \cos(k') R(\sin(k') - \sin(k)) \kappa(k'). \tag{C.12}$$

is needed. The second part can further be simplified by substituting $x = \sin(k') - \sin(k)$ and $dx = -\cos(k)dk$, yielding

$$\frac{\partial}{\partial k} \int_{-Q}^{Q} dk' \cos(k') R(\sin(k') - \sin(k)) \kappa(k')$$

$$= \int_{-Q}^{Q} dk' \cos(k') \frac{\partial}{\partial k} R(\sin(k') - \sin(k)) \kappa(k') \tag{C.13a}$$

$$= -\cos(k) \int_{-Q}^{Q} dk' \cos(k') R'(\sin(k') - \sin(k)) \kappa(k') \tag{C.13b}$$

with R' denoting the derivative $\frac{\partial R}{\partial x}$. By the use of this identity Eq. C.12 can be integrated by parts to obtain

$$\kappa'(k) = -\cos(k) \cos(k') R(\sin(k') - \sin(k)) \kappa(k') \Big|_{-Q}^{Q}$$
$$+ \cos(k) \int_{-Q}^{Q} dk' \cos(k') R(\sin(k') - \sin(k)) \frac{\partial \kappa(x(k'))}{\partial x}. \tag{C.14}$$

For a fixed μ the integration boundaries Q and $-Q$ determine the points where $\kappa(Q) = 0 = \kappa(-Q)$. Thus the first part of this equation vanishes, leaving

$$\kappa'(k) = \cos(k) \int_{x(-Q)}^{x(Q)} dx \frac{\cos(k')}{\cos(k')} R(x) \underbrace{\frac{\partial}{\partial x} \kappa(x(k'))}_{\frac{\partial \kappa(k')}{\partial k'} \frac{\partial k'}{\partial x}} \quad \text{(C.15a)}$$

$$= \cos(k) \int_{-Q}^{Q} R(\sin(k') - \sin(k)) \kappa'(k') dk'. \quad \text{(C.15b)}$$

In this way an integral equation for the derivative $\kappa'(k)$

$$\kappa'(k) = 2\sin(k) + \cos(k) \int_{-Q}^{Q} R(\sin(k') - \sin(k)) \kappa'(k') dk' \quad \text{(C.16)}$$

is obtained. This equation is again a Fredholm equation of the second kind and can be solved by the techniques explained above.

4. Using $\kappa'(k)$ and $\rho(k)$ the charge velocity obeys

$$v_c = \frac{\kappa'(k)}{p'(k)}\bigg|_{k=Q} \quad \text{(C.17a)}$$

$$= \frac{\kappa'(k)}{2\pi\rho(k)}\bigg|_{k=Q}. \quad \text{(C.17b)}$$

5. To determine the spin velocity the density function $\sigma_1(\Lambda)$ and the dressed energy $\epsilon_1(\Lambda)$ have to be determined. The density function σ_1 is given by a simple integral

$$\sigma_1(\Lambda) = \int_{-Q}^{Q} dk \frac{1}{4U} \frac{1}{\cosh(\frac{\pi}{2U}(\Lambda - \sin(k)))} \rho(k). \quad \text{(C.18)}$$

For large values of $\Lambda \to \infty$, which corresponds to the case of vanishing external field, the denominator can be expressed by

Determination of the Luttinger Parameters

the asymptotic behavior

$$\sigma_1(\Lambda) = \frac{1}{2U} e^{-\frac{\pi}{2U}\Lambda} \int_{-Q}^{Q} dk \, e^{\frac{\pi}{2U}\sin(k)} \rho(k). \quad \text{(C.19)}$$

6. In the same way the dressed energy ϵ_1 is given by an integral over κ

$$\epsilon_1(\Lambda) = \int_{-Q}^{Q} dk \frac{\cos(k)}{4U\cosh(\frac{\pi}{2U}(\Lambda-\sin(k)))} \kappa(k) \quad \text{(C.20)}$$

simplified by integration by parts to

$$\epsilon_1'(\Lambda) = \int_{-Q}^{Q} \frac{1}{4U\cosh(\frac{\pi}{2U}(\Lambda-\sin(k)))} \kappa'(k) dk. \quad \text{(C.21)}$$

For $\Lambda \to \infty$ Eq. C.21 can be simplified to

$$\epsilon_1'(\Lambda) = \frac{1}{2\pi} e^{-\frac{\pi}{2U}\Lambda} \int_{-Q}^{Q} e^{\frac{\pi}{2U}\sin(k)} \kappa'(k) dk. \quad \text{(C.22)}$$

7. With $\epsilon_1'(\Lambda)$ and $\sigma_1(\Lambda)$ the spin velocity is given as

$$v_s = \frac{\epsilon_1'(\infty)}{2\pi\sigma_1(\infty)}. \quad \text{(C.23)}$$

For $\Lambda = \infty$, i.e., in the case without external field, $\sigma_1(\Lambda)$ and $\epsilon_1(\Lambda)$ in Eq. C.18 and Eq. C.20 tend to zero. But in the expansion for large Λ both functions contain the same asymptotic factor. Thus the expanded versions of ϵ_1 and σ_1 given in Eq. C.19 and Eq. C.21 are used. In this way the exponential factors in the fraction cancel out

$$v_s = \frac{\int_{-Q}^{Q} e^{\frac{\pi}{2U}\sin(k)}dk}{2\pi \int_{-Q}^{Q} e^{\frac{\pi}{2U}\sin(k)}\rho(k)dk}. \quad (C.24)$$

Close to half-filling the charge velocity tends to zero as the charges are frozen out.

8. Now that the velocities are known, the last Luttinger parameter to be determined is K_ρ. The fourth parameter which has to be known is K_σ. But in the Hubbard model $K_\sigma = 1$ holds as the model is spin rotational invariant [89].
For the parameter K_ρ we have to set up the dressed charge matrix. Therefore the function $\xi(k)$ with

$$\xi(k) = 1 + \int_{-Q}^{Q} dk' \cos(k') R(\sin(k') - \sin(k)) \xi(k') \quad (C.25)$$

is determined by the Fredholm method. Finally the dressed charged matrix reads

$$Z = \begin{pmatrix} \xi(Q) & 0 \\ \frac{1}{2}\xi(Q) & \frac{1}{2}\sqrt{2} \end{pmatrix} \quad (C.26)$$

The anomalous dimension is then given by $\xi(Q)$.

Determination of the Luttinger Parameters

D. Comparison of the Two-Dimensional and the Infinite-Dimensional Model

Compared to the DMFT-data the results of the two-dimensional model reveal similar behavior for large interaction strengths U as expected from the discussion in Sect. 5.4.1.

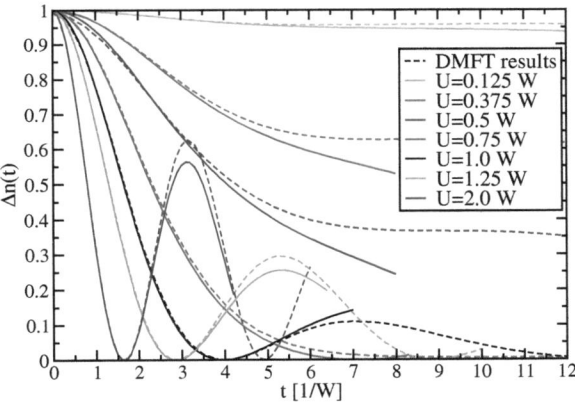

Fig. D.1.: Results for the jump in the two-dimensional model derived in a 9-loop calculation (solid lines) compared to the corresponding results obtained by DMFT [59] for the Bethe lattice with infinite coordination number (dashed lines).

For both approaches pronounced oscillations are found for strong quenches. The periods are comparable whereas the amplitudes of the oscillations differ (see Fig. D.1). For smaller quenches $U <$

Comparison of the Two-Dimensional and the Infinite-Dimensional Model

$0.75W$ the results for the two-dimensional model show faster relaxation than the ones for the Bethe lattice. This is probably due to the appearance of a plateau like feature in infinite dimensions, while the study of the two-dimensional model presented here does not show any sign of a plateau. In contrast to the DMFT-data the results of the equation of motion approach do not show any sign of a prethermalization plateau on the accessible time scales.

Bibliography

[1] M. Greiner, O. Mandel, T. Esslinger, T. W. Hänsch, and I. Bloch. *Nature* **415**, 39 (2002).

[2] T. Kinoshita, T. Wenger, and D. S. Weiss. *Nature* **440**, 900 (2006).

[3] L. Perfetti, P. A. Loukakos, M. Lisowski, U. Bovensiepen, H. Berger, S. Biermann, P. S. Cornaglia, A. Georges, and M. Wolf. *Phys. Rev. Lett.* **97**, 067402 (2006).

[4] I. Bloch, J. Dalibard, and W. Zwerger. *Rev. Mod. Phys.* **80**, 885 (2008).

[5] K. Morawetz, P. Lipavský, and M. Schreiber. *Phys. Rev. B* **72**, 233203 (2005).

[6] B.P. Anderson and M.A. Kasevich. *Science* **282**, 1686 (1998).

[7] R. Walters, G. Cotugno, T. H. Johnson, S. R. Clark, and D. Jaksch. *Phys. Rev. A* **87**, 043613 (2013).

[8] M. Greiner, O. Mandel, T. W. Hänsch, and I. Bloch. *Nature* **419**, 39 (2002).

[9] M. W. Zwierlein, C. H. Schunck, C. A. Stan, S. M. F. Raupach, and W. Ketterle. *Phys. Rev. Lett.* **94**, 180401 (2005).

[10] M. Greiner, C. A. Regal, and D. S. Jin. *Phys. Rev. Lett.* **94**, 070403 (2005).

[11] M. Köhl, H. Moritz, T. Stöferle, K. Günter, and T. Esslinger. *Phys. Rev. Lett.* **94**, 080403 (2005).

[12] J.K. Chin, D.E. Miller, Y. Liu, C. Stan, W. Setiawan, C. Sanner, K. Xu, and W. Ketterle. *Nature* **443**, 26 (2006).

Bibliography

[13] N. Strohmaier, D. Greif, R. Jördens, L. Tarruell, H. Moritz, T. Esslinger, R. Sensarma, D. Pekker, E. Altman, and E. Demler. *Phys. Rev. Lett.* **104**, 080401 (2010).

[14] F. Rossi and T. Kuhn. *Rev. Mod. Phys.* **74**, 895 (2002).

[15] V. Eisler and I. Peschel. *J. Stat. Mech.*, P06005 (2007).

[16] M. Diez, N. Chancellor, S. Haas, L. C. Venuti, and P. Zanardi. *Phys. Rev. A* **82**, 032113 (2010).

[17] T. Barthel and U. Schollwöck. *Phys. Rev. Lett.* **100**, 100601 (2008).

[18] M. Cramer, A. Flesch, I.P. McCulloch, U. Schollwöck, and J. Eisert. *Phys. Rev. Lett.* **101**, 063001 (2008).

[19] F. H. L. Essler, S. Evangelisti, and M. Fagotti. *Phys. Rev. Lett.* **109**, 247206 (2012).

[20] S. Sotiriadis and J. Cardy. *Phys. Rev. B* **81**, 134305 (2010).

[21] J. M. Deutsch. *Phys. Rev. A* **43**, 2046 (1991).

[22] M. Srednicki. *Phys. Rev. E* **50**, 888 (1994).

[23] M. Rigol, V. Dunjko, and M. Olshanii. *Nature* **452**, 854 (2008).

[24] R. Steinigeweg, J. Herbrych, and P. Prelovšek. *Phys. Rev. E* **87**, 012118 (2013).

[25] J.-S. Caux and F. H. L. Essler. *Phys. Rev. Lett.* **110**, 257203 (2013).

[26] S. R. Manmana, S. Wessel, R. M. Noack, and A. Muramatsu. *Phys. Rev. Lett.* **98**, 210405 (2007).

[27] G. Roux. *Phys. Rev. A* **79**, 021608 (2009).

[28] M. Rigol, V. Dunjko, V. Yurovsky, and M. Olshanii. *Phys. Rev. Lett.* **98**, 050405 (2007).

Bibliography

[29] M. Rigol. *Phys. Rev. Lett.* **103**, 100403 (2009).

[30] L. Amico and A. Osterloh. *J. Phys. A: Math. Gen.* **37**, 291 (2004).

[31] K. Sengupta, S. Powell, and S. Sachdev. *Phys. Rev. A* **69**, 053616 (2004).

[32] D. Rossini, A. Silva, G. Mussardo, and G.E. Santoro. *Phys. Rev. Lett.* **102**, 127204 (2009).

[33] G. Biroli, L. F. Cugliandolo, and A. Sicilia. *Phys. Rev. E* **81**, 050101 (2010).

[34] G. Roux. *Phys. Rev. A* **81**, 053604 (2010).

[35] E.T. Jaynes. *Phys. Rev.* **106**, 620 (1957).

[36] A. Iucci and M. A. Cazalilla. *Phys. Rev. A* **80**, 063619 (2009).

[37] M. A. Cazalilla. *Phys. Rev. Lett.* **97**, 156403 (2006).

[38] D. Fioretto and G. Mussardo. *New J. Phys.* **12**, 055015 (2010).

[39] P. Calabrese, F. H. L. Essler, and M. Fagotti. *Phys. Rev. Lett.* **106**, 227203 (2011).

[40] L. Foini, L. F. Cugliandolo, and A. Gambassi. *Phys. Rev. B* **84**, 212404 (2011).

[41] K. He, L. F. Santos, T. M. Wright, and M. Rigol. *Phys. Rev. A* **87**, 063637 (2013).

[42] J. Sirker, N.P. Konstantinidis, and N. Sedlmayr. *arXiv:* **1303.3064** (2013).

[43] J. Berges, S. Borsányi, and C. Wetterich. *Phys. Rev. Lett.* **93**, 142002 (2004).

[44] M. Eckstein, A. Hackl, S. Kehrein, M. Kollar, M. Moeckel, P. Werner, and F.A. Wolf. *Eur. Phys. J. Special Topics* **180**, 217 (2010).

Bibliography

[45] M. Eckstein, M. Kollar, and P. Werner. *Phys. Rev. Lett.* **103**, 056403 (2009).

[46] M. Moeckel and S. Kehrein. *Phys. Rev. Lett.* **100**, 175702 (2008).

[47] M. Moeckel and S. Kehrein. *Annals of Physics* **324**, 2146 (2009).

[48] M. Kollar, F. A. Wolf, and M. Eckstein. *Phys. Rev. B* **84**, 054304 (2011).

[49] A. Polkovnikov, K. Sengupta, A. Silva, and M. Vengalattore. *Rev. Mod. Phys.* **83**, 863 (2011).

[50] A. J. Daley, C. Kollath, U. Schollwöck, and G. Vidal. *J. Stat. Mech.* , P04005 (2004).

[51] S. R. White and A. E. Feiguin. *Phys. Rev. Lett.* **93**, 076401 (2004).

[52] G. Vidal. *Phys. Rev. Lett.* **98**, 070201 (2007).

[53] C. Kollath, A. M. Läuchli, and E. Altman. *Phys. Rev. Lett.* **98**, 180601 (2007).

[54] A. Yamamoto, M. Yamashita, and N. Kawakami. *J. Phys. Soc. Jpn.* **78**, 123002 (2009).

[55] F. B. Anders and A. Schiller. *Phys. Rev. Lett.* **95**, 196801 (2005).

[56] P. Schmidt and H. Monien. *cond-mat/0202046v1* (2002).

[57] J. K. Freericks, V. M. Turkowski, and V. Zlatic. *Phys. Rev. Lett.* **97**, 266408 (2006).

[58] A. Georges, G. Kotliar, W. Krauth, and M. J. Rozenberg. *Rev. Mod. Phys.* **68**, 13 (1996).

[59] M. Eckstein and M. Kollar. *Phys. Rev. Lett.* **100**, 120404 (2008).

Bibliography

[60] J. Sabio and S. Kehrein. *New J. Phys.* **12**, 055008 (2010).

[61] M. Heyl and S. Kehrein. *J. Phys.: Condens. Matter* **22**, 345604 (2010).

[62] G.G. Batrouni, F.F. Assaad, R.T. Scalettar, and P.J.H. Denteneer. *Phys. Rev. A* **72**, 031601(R) (2005).

[63] F. Goth and F. F. Assaad. *Phys. Rev. B* **85**, 085129 (2012).

[64] J. Hubbard. *Phys. Roy. Soc. Lond.* **276**, 238 (1963).

[65] M. C. Gutzwiller. *Phys. Rev. Lett.* **10**, 159 (1963).

[66] J. Kanamori. *Prog. Theor. Phys.* **30**, 275 (1963).

[67] F. Mott. *Metal-Insulator Transitions*. Taylor and Francis (1990).

[68] M. Imada, A. Fujimori, and Y. Tokura. *Rev. Mod. Phys.* **70**, 1039 (1998).

[69] J. G. Bednorz and K. A. Müller. *Zeitschrift fur Physik B Condensed Matter* **64**, 189 (1986).

[70] P. W. Anderson. *Science* **235**, 1196 (1987).

[71] E. H. Lieb and F. Y. Wu. *Phys. Rev. Lett.* **20**, 1445 (1968).

[72] G. S. Uhrig. *Phys. Rev. A* **80**, 061602(R) (2009).

[73] E. H. Lieb and D. W. Robinson. *Commun. math. Phys.* **28**, 251 (1972).

[74] P. Calabrese and J. Cardy. *Phys. Rev. Lett.* **96**, 136801 (2006).

[75] T. Enss and J. Sirker. *New J. Phys.* **95**, 023008 (2012).

[76] F. Iglói and H. Rieger. *Phys. Rev. Lett.* **85**, 3233 (2000).

[77] A.M. Läuchli and C. Kollath. *J. Stat. Mech.* **05**, P05018 (2008).

Bibliography

[78] S.R. Manmana, S. Wessel, R.M. Noack, and A. Muramatsu. *Phys. Rev. B* **79**, 155104 (2009).

[79] P. Roman. *Nuclear Physics B* **23**, 219 (1970).

[80] F.J. Wegner. *Lecture Notes* (2001).

[81] G. C. Wick. *Phys. Rev.* **80**, 268 (1950).

[82] J. Bonča, S. Maekawa, and T. Tohyama. *Phys. Rev. B* **76**, 035121 (2007).

[83] C. P. Heidbrink. Diploma thesis, Universität zu Köln, (2001).

[84] C. P. Heidbrink and G. S. Uhrig. *Eur. Phys. J. B* **30**, 443 (2002).

[85] A. Luther and I. Peschel. *Phys. Rev. B* **12**, 3908 (1975).

[86] F. D. M. Haldane. *Phys. Rev. Lett.* **45**, 1358 (1980).

[87] V. Meden and K. Schönhammer. *Phys. Rev. B* **46**, 15753 (1992).

[88] K. Penc and J. Sólyom. *Phys. Rev. B* **47**, 6273 (1993).

[89] J. Voit. *Rep. Prog. Phys.* **58** (1995).

[90] E. Miranda. *Braz. J. Phys.* **33**, 3 (2003).

[91] Giamarchi T. *Quantum Physics in One Dimension.* Oxford University Press, USA, (2004).

[92] A. Mitra and T. Giamarchi. *Phys. Rev. Lett.* **107**, 150602 (2011).

[93] J. Rentrop, D. Schuricht, and V. Meden. *New J. Phys.* **14**, 075001 (2012).

[94] D. M. Kennes and V. Meden. *arXiv:* **1304.5889** (2013).

[95] N. Nessi and A. Iucci. *Phys. Rev. B* **87**, 085137 (2013).

Bibliography

[96] C. Karrasch, J. Rentrop, D. Schuricht, and V. Meden. *Phys. Rev. Lett.* **109**, 126406 (2012).

[97] E. Coira, F. Becca, and A. Parola. *Eur. Phys. J. B* **86** (2013).

[98] F. Pollmann, M. Haque, and B. Dóra. *Phys. Rev. B* **87**, 041109 (2013).

[99] A. Iucci and M. A. Cazalilla. *New J. Phys.* **12**, 055019 (2010).

[100] C. De Grandi, V. Gritsev, and A. Polkovnikov. *Phys. Rev. B* **81**, 224301 (2010).

[101] A. Mitra and T. Giamarchi. *Phys. Rev. B* **85**, 075117 (2012).

[102] D. Iyer, H. Guan, and N. Andrei. *Phys. Rev. A* **87**, 053628 (2013).

[103] P. Jordan and E. Wigner. *Zeitschrift fur Physik* **47**, 631 (1928).

[104] E. Fradkin. *Field Theories of Condensed Matter Systems (Lecture Annote vol.82)*. Redwood City, CA: Addison-Wesley, (1991).

[105] F.D.M. Haldane. *Phys. Lett. A.* **81**, 153 (1981).

[106] F.D.M. Haldane. *J. Phys. C* **14**, 2585 (1981).

[107] C. N. Yang and C. P. Yang. *Phys. Rev.* **150**, 321 (1966).

[108] C. N. Yang and C. P. Yang. *Phys. Rev.* **150**, 327 (1966).

[109] U. Schollwöck. *Annals of Physics* **326**(1), 96 (2011).

[110] D. Peters, I. P. McCulloch, and W. Selke. *Phys. Rev. B* **85**, 054423 (2012).

[111] S. Tomonaga. *Progress of Theoretical Physics* **5**(4), 544 (1950).

[112] J. M. Luttinger. *J. Math. Phys.* **4**, 1154 (1963).

Bibliography

[113] K. Schönhammer. *J. Phys.: Condens. Matter* **25**, 014001 (2013).

[114] A. Luther and I. Peschel. *Phys. Rev. B* **9**, 2911 (1974).

[115] D.C. Mattis and E.H. Lieb. *J. Math. Phys.* **6**, 304 (1965).

[116] J. Voit. In *Electronic Properties Of Novel Materials-Molecular Nanostructures*, 544, 309. (2000).

[117] A. Theumann. *J. Stat. Phys.* **8**, 2460 (1967).

[118] J. von Delft and H. Schoeller. *Ann. Physik* **4**, 225 (1998).

[119] G. S. Uhrig. In *Correlated Fermionic Systems: Fermi Liquid and Luttinger Liquid*. http://t1.physik.tu-dortmund.de/uhrig/ (2005).

[120] J. Des Cloizeaux and M. Gaudin. *J. Math. Phys.* **7**, 1384 (1966).

[121] H. J. Schulz, G. Cuniberti, and P. Pieri. In *Field theories for low-dimensional condensed matter systems : spin systems and strongly correlated electrons,* G. Morandi, P. Sodano, A. Tagliacozzo, and V. Tognetti, editors. Springer(2000) (2000).

[122] J. Sólyom. *Phys. Rev. B* **28**, 201 (1979).

[123] Chen Z.-J., Zhang Y.-M., and Xu B.-W. *Commun. Theor. Phys.* **23**, 297 (1995).

[124] A. V. Chubukov, D. L. Maslov, and F. H. L. Essler. *Phys. Rev. B* **77**, 161102(R) (2008).

[125] R. Shankar. *Int. J. Mod. Phys. B* **4**, 2371 (1990).

[126] C. N. Yang and C. P. Yang. *Phys. Rev.* **147**, 303 (1966).

[127] J. Sirker and M. Bortz. *J. Stat. Mech.* **4**, P01007 (2006).

Bibliography

[128] H.J. Schulz. *Physica C: Superconductivity* **235, Part 1**(0), 217 (1994).

[129] F. H. L. Essler, H. Frahm, F. Göhmann, A. Klümper, and V. E. Korepin. *The One-Dimensional Hubbard Model*. Cambridge University Press (2005).

[130] J. Voit. *Int. J. Mod. Phys. B* **5**, 8305 (1993).

[131] Johannes Voit. *Phys. Rev. B* **47**, 6740 (1993).

[132] H. J. Schulz. *Phys. Rev. Lett.* **64**, 2831 (1990).

[133] H. Frahm and V. E. Korepin. *Phys. Rev. B* **42**, 10553 (1990).

[134] F. Göhmann and V. E. Korepin. *Physics Letters A* **263**, 293 (1999).

[135] V. J. Emery. In *Presented at the 2nd Summer Inst. on Phys. of Low Dimensional Systems, Kyoto, Japan, 7, 1979*, 7. (1979).

[136] M. Tsuchiizu and A. Furusaki. *Phys. Rev. B* **69**, 035103 (2004).

[137] J. Voit. *Phys. Rev. B* **45**, 4027 (1992).

[138] W. Metzner and D. Vollhardt. *Phys. Rev. B* **39**, 4462 (1989).

[139] S. A. Hamerla and G. S. Uhrig. *New J. Phys.* **15**, 073012 (2013).

[140] E. Müller-Hartmann. *Z. Phys. B* **74**, 507 (1989).

[141] B. Sciolla and G. Biroli. *Phys. Rev. Lett.* **105**, 220401 (2010).

[142] B. Sciolla and G. Biroli. *J. Stat. Mech.* , P11003 (2011).

[143] M. Schiró and M. Fabrizio. *Phys. Rev. Lett.* **105**, 076401 (2010).

[144] M. Sandri, M. Schiró, and M. Fabrizio. *Phys. Rev. B* **86**, 075122 (2012).

Bibliography

[145] M. Schiró and M. Fabrizio. *Phys. Rev. B* **83**, 165105 (2011).

[146] M. Abramowitz and I. A. Stegun. *Handbook of Mathematical Functions: with Formulas, Graphs, and Mathematical Tables.* Dover Books on Mathematics, (1972).

[147] W. F. Brinkman and T. M. Rice. *Phys. Rev. B* **2**, 1324 (1970).

[148] Walter Metzner and Dieter Vollhardt. *Phys. Rev. B* **37**, 7382 (1988).

[149] F. Gebhard. *Phys. Rev. B* **41**, 9452 (1990).

[150] S. A. Hamerla and G. S. Uhrig. *Phys. Rev. B* **87**, 064304 (2013).

[151] E.J. Torres-Herrera and L.F. Santos. *arXiv:* **1305.6937** (2013).

[152] M. Mierzejewski, L. Vidmar, J. Bonča, and P. Prelovšek. *Phys. Rev. Lett.* **106**, 196401 (2011).

[153] J. Bonča, M. Mierzejewski, and L. Vidmar. *Phys. Rev. Lett.* **109**, 156404 (2012).

[154] G. Czycholl. *Theoretische Festkörperphysik.* Springer Berlin Heidelberg (2004).

[155] A. Auerbach. *Interacting Electrons and Quantum Magnetism.* Springer New York, (1994).

[156] S. A. Hamerla and G. S. Uhrig. *arXiv:* **1307.3438** (2013).

[157] L. Erdös, M. Salmhofer, and H.-T. Yau. *J. Stat. Phys.* **116**, 367 (2004).

[158] M. Kollar. *private communications* .

[159] H.-P. Breuer and F. Petruccione. *The Theory of Open Quantum Systems.* Clarendon Press, Oxford (2006).

[160] S. Blanes, F. Casas, J. A. Oteo, and J. Ros. *Phys. Rep.* **470**, 151 (2009).

Bibliography

[161] N. Tsuji, P. Barmettler, H. Aoki, and P. Werner. *arXiv:* **1307.5946** (2013).

[162] W. H. Press, S. A. Teukolsky, W. T. Vetterling, and B. P. Flannery. *Numerical Recipes in* C++. Cambridge University Press, (2002).

Teilpublikationen

Teile der vorliegenden Arbeit wurden an folgenden Stellen veröffentlicht:

1. Simone A. Hamerla und Götz S. Uhrig
   Dynamical transition in interaction quenches of the one-dimensional Hubbard model
   Phys. Rev. B 87, 064304 (2013)

2. Simone A. Hamerla und Götz S. Uhrig
   One-dimensional fermionic systems after interaction quenches and their description by bosonic field theories
   New J. Phys. 15, 073012 (2013)

3. Simone A. Hamerla und Götz S. Uhrig
   Interaction quenches in the two-dimensional fermionic Hubbard model
   arXiv: 1307.3438 (2013)

Bibliography

Danksagung

An dieser Stelle möchte ich mich bei den Menschen bedanken, die mich auf meinem Weg unterstützt haben.
Zunächst möchte ich mich bei Prof. Dr. Götz S. Uhrig für die Vergabe dieses spannenden Themas und die hervorragende Betreuung bedanken. Bei Prof. Dr. Frithjof Anders und Prof. Dr. Alejandro Muramatsu möchte ich mich herzlich für die Begutachtung dieser Arbeit bedanken.
Des Weiteren gilt mein Dank der Studienstiftung des deutschen Volkes für die finanzielle Unterstützung und die vielen Unternehmungen. Der Japan Society for the Promotion of Science danke ich für die finanzielle Unterstützung während meines Japan-Aufenthaltes - es war eine wunderbare Zeit.
Dr. Marcus Kollar, Prof. Dr. Martin Eckstein, Prof. Dr. Volker Meden, Prof. Dr. Philipp Werner und Naoto Tsuji danke ich für die Bereitstellung der DMFT- und der DCA-Daten und natürlich für die zahlreichen Diskussionen und Anregungen.
Gregor Foltin und Daniel Stanek möchte ich für die herzliche und auch lustige Atmosphäre im Büro danken. Carsten Raas danke ich besonders für die Unterstützung im Kampf gegen/mit dem Rechner. Außerdem danke ich Dr. Fabian Güttge und Björn Zelinski für die vielen spaßigen Momente. Meinen Kollegen und Freunden danke ich für die schöne Zeit.
Prof. Dr. Joachim Stolze danke ich dafür, dass er immer ein offenes Ohr und gute Ratschläge für mich hat.
Dr. Thomas Schmidt gilt mein besonderer Dank für das Korrekturlesen der vorliegenden Arbeit.
Besonders möchte ich mich bei Dr. Isabelle Exius und Claudia Zens bedanken, dafür dass sie immer für mich da sind und mir beistehen wenn ich sie brauche. Vielen Dank dafür.

Zu guter Letzt möchte ich mich besonders bei meiner Familie bedanken, die immer hinter mir steht und immer an mich geglaubt hat. Ich danke meinen Eltern, den liebevollsten Menschen die ich kenne,

Bibliography

für ihre unendliche Liebe und Unterstützung in jeglicher Hinsicht. Meinem Bruder gilt besonderer Dank dafür, dass er stets hinter mir steht, ohne mich dabei vergessen zu lassen, dass ich bei uns beiden immer den Kürzeren ziehen werde.

i want morebooks!

Buy your books fast and straightforward online - at one of world's fastest growing online book stores! Environmentally sound due to Print-on-Demand technologies.

Buy your books online at
www.get-morebooks.com

Kaufen Sie Ihre Bücher schnell und unkompliziert online – auf einer der am schnellsten wachsenden Buchhandelsplattformen weltweit! Dank Print-On-Demand umwelt- und ressourcenschonend produziert.

Bücher schneller online kaufen
www.morebooks.de

 VDM Verlagsservicegesellschaft mbH
Heinrich-Böcking-Str. 6-8   Telefon: +49 681 3720 174   info@vdm-vsg.de
D - 66121 Saarbrücken   Telefax: +49 681 3720 1749   www.vdm-vsg.de

Printed by Books on Demand GmbH, Norderstedt / Germany